Microbial and Phenotypic Definition of Rats and Mice
Proceedings of the 1998 US/Japan Conference

International Committee of the Institute for Laboratory Animal Research
National Research Council

NATIONAL ACADEMY PRESS
Washington, D.C.

NATIONAL ACADEMY PRESS • **2101 Constitution Avenue, NW** • **Washington, DC 20418**

NOTICE: The project that is the subject of this report was approved by the Governing Board of the National Research Council, whose members are drawn from the councils of the National Academy of Sciences, the National Academy of Engineering, and the Institute of Medicine. The members of the committee responsible for the report were chosen for their special competences and with regard for appropriate balance.

This study was supported by Grant No. P40-RR-11611 between the National Academy of Sciences and the National Institutes of Health. Any opinions, findings, conclusions, or recommendations expressed in this publication are those of the author(s) and do not necessarily reflect the views of the organizations or agencies that provided support for the project.

International Standard Book Number 0-309-06591-7

Copyright 1999 by the National Academy of Sciences. All rights reserved.

Printed in the United States of America.

INTERNATIONAL COMMITTEE OF THE INSTITUTE FOR LABORATORY ANIMAL RESEARCH COUNCIL

Christian R. Abee (*Chair*), Department of Comparative Medicine, University of South Alabama, Mobile, AL
Hilton J. Klein, Department of Laboratory Animal Resources, Merck Research Laboratories, West Point, PA
William Morton, Regional Primate Research Center, University of Washington, Seattle, WA
Robert J. Russell, Harlan Sprague Dawley, Inc., Indianapolis, IN
William S. Stokes, Environmental Toxicology Program, National Institute of Environmental Health Sciences, Research Triangle Park, NC
John L. VandeBerg, Southwest Foundation for Biomedical Research, San Antonio, TX
Peter A. Ward, Department of Pathology, University of Michigan Medical School, Ann Arbor, MI

Staff

Ralph B. Dell, Director
Kathleen A. Beil, Administrative Assistant
Susan S. Vaupel, Managing Editor, *ILAR Journal*
Marsha K. Williams, Project Assistant

The National Academy of Sciences is a private, nonprofit, self-perpetuating society of distinguished scholars engaged in scientific and engineering research, dedicated to the furtherance of science and technology and to their use for the general welfare. Upon the authority of the charter granted to it by the Congress in 1863, the Academy has a mandate that requires it to advise the federal government on scientific and technical matters. Dr. Bruce M. Alberts is president of the National Academy of Sciences.

The National Academy of Engineering was established in 1964, under the charter of the National Academy of Sciences, as a parallel organization of outstanding engineers. It is autonomous in its administration and in the selection of its members, sharing with the National Academy of Sciences the responsibility for advising the federal government. The National Academy of Engineering also sponsors engineering programs aimed at meeting national needs, encourages education and research, and recognizes the superior achievements of engineers. Dr. William A. Wulf is president of the National Academy of Engineering.

The Institute of Medicine was established in 1970 by the National Academy of Sciences to secure the services of eminent members of appropriate professions in the examination of policy matters pertaining to the health of the public. The Institute acts under the responsibility given to the National Academy of Sciences by its congressional charter to be an adviser to the federal government and, upon its own initiative, to identify issues of medical care, research, and education. Dr. Kenneth I. Shine is president of the Institute of Medicine.

The National Research Council was organized by the National Academy of Sciences in 1916 to associate the broad community of science and technology with the Academy's purposes of furthering knowledge and advising the federal government. Functioning in accordance with general policies determined by the Academy, the Council has become the principal operating agency of both the National Academy of Sciences and the National Academy of Engineering in providing services to the government, the public, and the scientific and engineering communities. The Council is administered jointly by both Academies and the Institute of Medicine. Dr. Bruce M. Alberts and Dr. William A. Wulf are chairman and vice chairman, respectively, of the National Research Council.

Preface

US-Japan meetings on laboratory animal science have been held virtually every year since 1980 under the US-Japan Cooperative Program on Science and Technology. Over the years these meetings have resulted in a number of important documents including the *Manual of Microbiologic Monitoring of Laboratory Animals* published in 1994 and the article *Establishment and Preservation of Reference Inbred Strains of Rats for General Purposes* published in 1991. In addition to these publications, these meetings have been instrumental in increasing awareness of the need for microbiologic monitoring of laboratory rodents and the need for genetic definition and monitoring of mice and rats.

In cooperation with the Comparative Medicine section of NCRR/NIH, ILAR Council and staff are pleased to become the host for this important annual meeting and look forward to participating in future meetings. The support and sponsorship of NCRR (P40 RR 11611) in the United States and the Central Institute for Experimental Animals in Japan are gratefully acknowledged. These meetings have increased understanding of American and Japanese approaches to laboratory animal science and should continue to strengthen efforts to harmonize approaches aimed at resolving common challenges in the use of animal models for biomedical research and testing. This effort to improve understanding and cooperation between Japan and the United States should also be useful in developing similar interaction with other regions of the world including Europe, Australia, and Southeast Asia.

Christian R. Abee, *Chair*
International Committee of the Institute
for Laboratory Animal Research

Contents

Opening Remarks   1
   Shin-ichi Ota
      Historical Overview, 1
      Expectations, 2
      References, 2

Opening Remarks   3
   Judith L. Vaitukaitis

The Need for Defined Rats and Mice in Biomedical Research:
Problems, Issues, and the Current State of Affairs   5
   Tatsuji Nomura
      Introduction, 5
      Participants and Format, 5
      Standardization and Quality, 6
      Questions and Answers, 6

The Need for Defined Rats and Mice in Biomedical Research:
Problems, Issues, and the Current State of Affairs   7
   Norikazu Tamaoki
      Global Health Issue and the Necessity of Laboratory Animals, 7
      Importance of Laboratory Animals for Human Health, 7
      Laboratory Animal Models in Major Disease Categories, 8
      Aspects of Future Laboratory Animal Use, 10
      Questions and Answers, 10
      References, 11

The Biological Integrity of Laboratory Rodents   12
    Robert O. Jacoby
        References, 14

Quality Testing System for SPF Animals in Japan and Problems in the
Management of Such Systems   15
    Toshio Itoh
        Role of the ICLAS Monitoring Center in the Quality Control
            System of Laboratory Animals, 15
        Microbiological Monitoring System of the ICLAS Monitoring
            Center, 16
        Microbiological Contamination of Laboratory Animals in Japan, 17
        Conclusion, 22
        References, 23

Definition of Microbiological Status of Rats and Mice /
The Need for Methods of Defining Flora /
International Standards for Terminology   24
    Kazuaki Mannen
        Current Status of Microbiological Quality of Laboratory
            Animals in University Animal Centers in Japan, 24
        Importance of Identifying Contamination, 25

Development of Rodent Pathogen Profiles and Adequacy of Detection
Technology   28
    Steven H. Weisbroth
        Problematic Issues, 36
        References, 38

Current Status of Pathogen Status in Mice and Rats   39
    J. Russell Lindsey
        Past Progress toward Reducing Pathogens, 39
        Pathogens Still Pose Pervasive Risks in the United States, 40
        Pathogen Status Gets Lost in the Terminology Morass, 40
        Principles of Pathogen Status Are Being Compromised, 41
        Scientists Have Little Appreciation of Pathogen Status, 41
        References, 42

Genetic Background and Phenotypes in Animal Models of Human Diseases 44
    Kuzuo Moriwaki
        Development of Experimental Mouse Strains, 44
        Transgenic and Knockout Mice, 45
        Recombinant Inbred Strains, 45
        Common Disease Models, 46
        Conclusion, 46
        References, 47

CONTENTS

Genetic and Phenotypic Definition of Laboratory Mice and Rats /
What Constitutes an Acceptable Genetic-Phenotypic Definition   48
    Hideki Katoh
        Genetic Studies on Closed Colonies of the Rat, 48
        Human Ancestry, 48
        Laboratory Animals, 51
        Demonstration of the Existence of Closed Colonies Using
            Genetic Monitoring, 52
        Summary and Discussion, 54
        References, 57

Phenotype Assessment Requires More Than a Casual Observation   58
    Philip A. Wood
        Primary Level Assessment: Find Abnormalities, 59
        Secondary Level Assessment: Evaluate and Quantify
            Abnormalities, 59
        Environmental Influences, 61
        Examples, 61
        References, 62

Genetic and Phenotypic Definition of Laboratory Mice and Rats /
What Constitutes an Acceptable Genetic-Phenotypic Definition   63
    Muriel T. Davisson
        Genetically Defined Mice, 63
        Genetic Standardization, 63
        Genetic Nomenclature, 64
        Definition and Value of Different Kinds of Strains, 65
        Genetic Monitoring, 67
        Genetic Databases, 68
        Training Scientists to Use Genetically Defined Mice, 70
        References, 70

Genetic and Phenotypic Definition of Laboratory Mice and Rats /
What Constitutes an Acceptable Genetic-Phenotypic Definition   71
    Joseph DeGeorge
        Necessity for Globally Standardized Outbred Rats for
            Carcinogenicity Bioassay, 71
        Change in Laboratory Animal Science, 71
        Globalization of Drug Development, 72
        Segmentation of Toxicology Testing, 72
        Need for Integrated Findings, 74
        Managing Changes Over Time, 74

CIEA/NCRR/NIH Genetic and Microbiological Monitoring of Mouse and
Rat Resources: Directions for the Future   76
   Tatsuji Nomura
      Differences Between Countries, 76
      Questions and Answers, 77

CIEA/NCRR/NIH Genetic and Microbiological Monitoring of
Mouse and Rat Resources: Directions for the Future   78
   Neal West
      NIH Structure, 78
      NCRR's Mission, 78
      Database Recommendation, 80
      Questions and Answers, 80
      Learning from Each Other, 81
      Policy Setting, 81
      Reference, 82

Closing Comments / Summary of Presentations   83
   Thomas J. Gill III
      Introduction, 83
      Need for Genetically Defined Animals, 83
      Development of Genetically Engineered Animals, 84
      Importance of Disease Models, 86
      Standardization and Monitoring, 87
      Rat Repository Workshop, 87
      National Rat Genetics Resources Center, 88
      Recommendations, 89

Summary of Presentations   90
   Steven P. Pakes
      Current Status of Laboratory Animal Science, 90
      Revitalization of Original Focus, 90
      Future Cooperation, 91

Summary of Presentations   92
   Tatsuji Nomura
      Laboratory Animal Science: 1950s to 1990s, 92
      Genetically Engineered Animals, 93
      Conclusion, 94

Appendix A: US/Japan Meeting Agenda   95

Appendix B: Meeting Participants   98

Opening Remarks

Judith L. Vaitukaitis
Director, National Center for Research Resources
National Institutes of Health
Bethesda, Maryland

It is fitting that this meeting is taking place today at the National Academy of Sciences. Under this roof, the National Research Council—founded in 1916 under President Woodrow Wilson—has worked diligently for the betterment of science through cooperation among public and private as well as national and international organizations. To this purpose, cooperation and harmony knit a smaller world where scientific interaction and cooperation grows larger—crossing not only the boundaries of research disciplines, but those of countries and continents as well. We monitor trends in biomedical research, set priorities, and focus the research community. As a result, we markedly expand our insights and create opportunities for the promise of improving world health.

The governments of the United States and Japan have long recognized the importance of scientific cooperation and the synergy that it generates. A joint program established nearly two decades ago by these governments has spawned new projects that have helped to advance techniques and establish standards for biomedical research around the world. The National Institutes of Health and the Japanese Central Institute for Experimental Animals, under the terms of this program, have long nurtured collaborative, information-exchange activities. The tenacity of many leaders—including Drs. Nomura, Kagiuama, Held, and Allen as well as other important contributors—has immensely enhanced the genetic and microbiologic integrity of laboratory rat and mouse colonies, not only in the United States and Japan, but worldwide. Advanced microbiologic monitoring for major infectious agents and improved diagnostic techniques for diseases now safeguard our valuable but fragile resource investments, including specific pathogen-free animals.

Through improved monitoring and sophisticated husbandry, the opportunities for biomedical investigators to create and use unique and complex animal models have multiplied, resulting in powerful research tools. As genetics and genomics rapidly and dramatically affect the study of biology and medicine, the role of comparative medicine cannot be understated. Not long ago, the capacity to remove or alter with precision a single gene among many thousands in the genome of an animal and to transmit this mutation to all subsequent progeny was considered nothing short of science fiction. Now, as a result of such revolutionary breakthroughs, investigators interested in understanding the structure and function of specific genes and their expressed macromolecules are demanding new cutting-edge research resources and technologies. They are asking for sophisticated, high-quality animal models; new and advanced instruments; and technologies.

The mouse has been a critical model for the identification of brain lesions in Huntington's disease, the discovery of genes responsible for several cancers, and many other diseases. The rat model—although not currently as robust as the mouse in many ways—is, however, the best "functionally" characterized mammalian model system. As production of transgenic rats becomes routine in many laboratories, including commercial settings, well-characterized, genetically altered rat models will contribute significantly to studies of human biology and disease. Mouse and rat model systems will further enable investigators to discover gene function by linking physiology, genetics, and clinical phenotypes.

Today's meeting is critical to helping the global scientific enterprise harmonize the mouse and rat models and to meeting research resource challenges of the 21st century. Over the years, you have shared your insights and have challenged conventional wisdom about laboratory animal sciences; you have also expressed your vision for the future and then set the wheels in motion for new generations of molecular biologists. Together our strong commitment to laboratory animal research infrastructure will pave the way for further refining these valuable genetic resources. I look forward to your fruitful discussions and insights.

The Need for Defined Rats and Mice in Biomedical Research: Problems, Issues, and the Current State of Affairs

Tatsuji Nomura
Director, Central Institute for Experimental Animals
Kawasaki, Japan

INTRODUCTION

Over the years, discussions held during the US/Japan Meetings have provided a technical basis for the genetic and microbiological testing conducted by the International Council for Laboratory Animal Science (ICLAS) monitoring center. They also have made a major contribution toward establishing the concept of quality standards of laboratory animals on an international level.

PARTICIPANTS AND FORMAT

Originally the two main participants in this meeting were the Veterinary Resources Branch (VRB), Division of Research Services (DRS), at the National Institutes of Health (NIH) on the US side and the Central Institute for Experimental Animals (CIEA) on the Japanese side. At present, the original US participant has been replaced by the Comparative Medicine Program of the NIH National Center for Research Resources (NCRR) with assistance from the Institute for Laboratory Animal Research (ILAR), National Research Council (NRC), National Academy of Sciences (NAS). The basic concept of the US/Japan Meeting has always been an exchange of experience and technology concerning important problems related to laboratory animals of mutual interest to both countries. In 1996, Dr. Leo Whitehair of the NCRR assumed responsibility in place of the VRB, and the meeting was given a new start under a new format with ILAR Director Dr. Ralph Dell participating.

This 19th US/Japan Meeting is the first meeting under the new format. The

Japanese side would like to propose that quality control of laboratory animals continue as the main topic in the future under the basic concept of the US/Japan Cooperative Program on Science and Technology. This year the topic will be quality standards. Recently animals introduced into Japan from overseas have been the cause of microbiological contamination, and we want to discuss this problem under the topic of microbiological quality.

STANDARDIZATION AND QUALITY

With respect to genetic quality, the International Conference of Harmonization of Technical Requirements for International Registration for Pharmaceuticals for Human Use (ICH) has decided that one set of animal experimentation data will be used in all countries when applying for new drug approvals. However, global standardization of rat closed colony stock and a genetic monitoring system for use in 2-year carcinogenicity bioassays has still not been established. Therefore, these problems must be solved as soon as possible.

Quality control has become a significant international issue, and I hope we will have a fruitful discussion on this and other issues related to laboratory animal science of interest to the two countries. Among our topics of discussion will be global health issues of experimental animals and the need for defining laboratory animals.

CIEA is supported financially by the Ministry of Education, Science, and Culture because the monitoring center undertakes genetic and microbiological monitoring for universities under the control of the Ministry. In addition, CIEA has 44 supporting members from industry that pay an annual fee and also give donations because toxicology is one of the most critical animal studies for the pharmaceutical industry. Their requirements are very strict, which has increased the level of animal experimentation.

Laboratory animal science is a very broad field, covering many disciplines, and it requires good collaboration between industry and academia. From the Japanese view, I wonder why the United States pharmaceutical industry does not support laboratory animal science. I also would like to know where the laboratory animal centers are and who are the US opinion leaders.

QUESTIONS AND ANSWERS

T. GILL: Certainly there is a large number of organisms that can be monitored and genes that can be tested. I believe the critical issue is how to select the microorganisms or genes. A second category is local problems, some of which must also be monitored. I believe this group should generate recommendations about what is essenial and what are special local needs.

The Need for Defined Rats and Mice in Biomedical Research: Problems, Issues, and the Current State of Affairs

Norikazu Tamaoki
Professor, Department of Pathology, Tokai University School of Medicine
Kanagawa, Japan

GLOBAL HEALTH ISSUE AND THE NECESSITY OF LABORATORY ANIMALS

Speaking on behalf of the Liaison Committee for Laboratory Animal Science of the Science Council of Japan, I am pleased to discuss laboratory animals from the viewpoint of global health issues. My talk will be rather general and will include the following three major topics: (1) importance of laboratory animals for human health, (2) laboratory animal models in major disease categories, and (3) aspects of future laboratory animal use.

IMPORTANCE OF LABORATORY ANIMALS FOR HUMAN HEALTH

Many aspects of health issues are directly related to the socioeconomic status of the world's regions and countries. In the populations of developing countries, nutrition and infection are urgent problems to be solved. However, in the developed countries, life-style diseases are important issues. Emerging and reemerging infectious diseases and drug abuse are important in both developing and developed countries.

Risk factor analysis shows that 50% of all types of disease is due to life-style, 20% to environmental factors, 20% to genetic factors, and the remaining 10% to medical care. To prevent and treat such illnesses, it is necessary to understand the mechanism of diseases and to develop intervention systems including medical care. New drugs and health education are indispensable for disease prevention.

For this purpose, we need more accurate models and assay systems for both normal and abnormal human conditions.

Recent results of experimental gene therapy against tumor angiogenesis serve as an important example of the necessity of a whole organism model. It has been documented that growth of human cancer in vivo is dependent on vascularity and blood supply. We have shown that one of five isoforms of vascular endothelial growth factor—VEGF 189, a potent vascular growth factor—is responsible for growth and metastases of human cancers including colon (Tokunaga and others 1998), lung (Oshika and others 1998), and kidney (Tomisawa and others 1999). As a model for cancer gene therapy, transfection of the ribozyme that specifically catalyzes VEGF 189 into a cancer cell line has little effect on cell growth in vitro. In contrast, the same procedure inhibits in vivo human tumor growth xenotransplanted in severe combined immunodeficiency disorders (SCID) mice by suppressing angiogenesis (Oshika and Nakamura, manuscript in preparation). Clearly, to study the complex function of the organism's multicellular or multi-organ system requires use of a whole organism (namely an animal) model.

Accumulating data on human and mouse genomes and advances in gene technology have enabled us to have a more accurate understanding of gene structure and function. However, with preliminary results obtained from genetically engineered mice, we are still far from our goal of understanding the function of the whole organism. Knockout or transgenic mice that have been developed represent the change in only one or a few among 100,000 genes in the whole genome. Gene function is not uniformly expressed in cases involving alternative splicing or other mechanisms resulting in production of several isoforms of gene products with different biological activity. In addition, evaluation of a cancer gene therapy model requires not only cancer cells, but also supporting tissue.

LABORATORY ANIMAL MODELS IN MAJOR DISEASE CATEGORIES

Major disease categories for which appropriate animal models are needed include infectious disease, immunological disease, cancer, and life-style diseases. Details about these disease categories follow.

Infectious Disease

Elucidation of receptor molecules for microorganisms and toxins enables us to change the ordinary host range and to develop animal models susceptible to various human-specific pathogens. Polio virus-susceptible mice produced by CIEA are a good example. Additional studies of virus receptors and coreceptors will hopefully create various models for human infectious diseases. Animal models for parasitic disease is very important, but few practical models exist.

Recent studies on cytokine reaction patterns disclosed TH1- and TH2-type immune response. These results are helpful for studying the mechanisms involved in leishmaniasis and schistosomiasis (Mosman and Coffman 1989; Wynn and others 1995). In addition, progress in the development of new vaccinations such as DNA vaccine has been made through studies with laboratory animals.

Immunological Disease

Many types of immune-deficient models have been useful for human to mouse tissue xenotransplants. SCID mice and Rag2 KO mice are also interesting from the standpoint of genetic instability. Autoimmune models have a long history starting from the study of mouse genetics in mutants. Allergy models have been developed in many institutions. In addition, the study of cell adhesion molecules in transgenic and knockout mice has stimulated progress in the study of inflammation and in understanding cell behavior in vivo.

Cancer

Genetically engineered mice have played an important role in cancer research results based on the accumulation of data related to prenatal and postnatal gene abnormalities. Application of these data to angiogenicity assays appears to be a very promising biomedical tool for cancer treatment and prevention. The effects of background genes are also very important for our understanding of metabolism of carcinogens and organ-specific development of tumors.

Life-style Diseases

The life-style disease category includes major diseases in developed countries, such as cardiovascular disease and diabetes. Considerable progress has been made in the field of cardiovascular disease, including hypertension and atherosclerosis, using transgenic mice expressing human renin-angiotensin gene and scavenger receptor gene. Here again, the whole body model is very important in understanding pathophysiology of disease due to multiple gene errors superimposed on human life-style factors.

Risk Assessment for Environmental Factors

Laboratory animals are important for the assessment of environmental risk because they serve as whole organism models for many risks of unknown etiology. The models are used not only for drug testing, but also for assay systems to evaluate various risks to human health. Selecting appropriate models for specifically targeted risks is extremely important.

ASPECTS OF FUTURE LABORATORY ANIMAL USE

For our common goal—human health—we must develop and produce objectively oriented and quality-controlled laboratory animals. Laboratory animals used in the future should be purposefully selected for particular studies, be of reasonable cost both monetarily and in terms of life expectancy and maintenance, and be easily available. Although we currently have many disease models for biomedical research, most of the animals used for specific purposes do not fulfill the criteria described above.

We need to develop a specialized support system to supply future laboratory animals. Such a system should be based on interdisciplinary research (including gene technology), biological databases, the entire field of human health science, novel methods of reproduction and development, and a system of animal care based on laboratory animal science.

In conclusion, I would like to propose the establishment of new collaborative networks to unite government, academia, and industry for the development of the future of laboratory animals and laboratory animal science.

QUESTIONS AND ANSWERS

C. ABEE: With regard to your comment about the collaboration of government, industry, and academia, could you please provide your perspective on how that is done in Japan?

N. TAMAOKI: I think there is no such system in Japan. Individual institutes such as CIEA conduct it; however, as Dr. Gill pointed out, laboratory animal technology is progressing rapidly and it is too expensive to keep them in one institute. I believe that we need more united resources from governmental budgets and industry monies to maintain standardized, high-quality animals. I would also like to mention the pitfalls that exist with laboratory animals as a result of transgenics. There is a great difference between such genetically engineered animals and the reliability and availability of other laboratory animals. For this reason, we need a bridge between research models and laboratory animals, which ideally would be supported by government and industry. If Japanese and US governments collaborate on this point, it would be better for both countries. Unfortunately, we do not have a real system at this time.

The most important issue right now is how to define the phenotype of animals. It is very important to bridge the gap between genotype and phenotype. In the research field, it is very important to study the expression and mechanisms of genes. Gene expression is controlled in gene products of other genes. There are networks or cascades of functional products of genes, but at present, we do not have enough methods to check such a process, which takes time. For the moment, I think the practical way is to create a new method for defining the phenotype of animals. The functional phenotype must be defined by the reaction of animals to some standard substance depending on the objectives of the experiment—for a metabolism study, for a neurology study, and so on.

REFERENCES

Mosman, T. R., and R. L. Coffman. 1989. TH1 and TH2 cells: Different patterns of lymphokine secretion lead to different functional properties. Ann. Rev. Immunol. 7:145-173.

Oshika, Y., M. Nakamura, T. Tokunaga, Y. Ozeki, Y. Fukushima, H. Hatanaka, Y. Abe, H. Yamazaki, H. Kijima, N. Tamaoki, and Y. Ueyama. 1998. Expression of cell-associated isoform of vascular endothelial growth factor 189 and its prognostic relevance in non-small cell lung cancer. Int. J. Oncol. 12:541-544.

Tokunaga, T., Y. Oshika, Y. Abe, Y. Ozeki, S. Sadahiro, H. Kijima, T. Tsuchida, Y. Yamazaki, Y. Ueyama, N. Tamaoki, and M. Nakamura. 1998. Vascular endothelial growth factor (VEGF) mRNA isoform expression pattern is correlated with liver metastasis and poor prognosis in colon cancer. Br. J. Cancer 77:998-1002.

Tomisawa, M., T. Tokunaga, Y. Oshika, T. Tsuchida, Y. Fukushima, H. Sato, H. Kijima, H. Yamazaki, Y. Ueyama, N. Tamaoki, and M. Nakamura. 1999. Expression pattern of vascular endothelial growth factor isoform is closely correlated with tumour stage and vascularization in renal cell carcinoma. Eur. J. Cancer 35:133-137.

Wynn, T. H., D. Jankovic, S. Hiney, A. W. Cheever, and A. Sher. 1995. IL-12 enhances vaccine-induced immunity to schistosomiasis mansoni in mice and decreases T helper 2 cytokine expression, IgE production, and tissue eosinophilia. J. Immunol. 154:4701-4709.

The Biological Integrity of Laboratory Rodents

Robert O. Jacoby
Professor of Comparative Medicine, Yale University
New Haven, Connecticut

The value of small rodents to biomedical research is beyond question. About half of the extramural grants awarded annually by the US National Institutes of Health require some use of laboratory animals. Because more than 95% of such use involves mice and rats, public investment in rodent-based research is enormous. Statistics for research in Japan are probably comparable. It follows that the quality of much biomedical research relies substantively on the quality of laboratory rodents. During 1993, the NCRR conducted a national survey of laboratory animal use, facilities, and resources. It reported that more than 7,000,000 mice were used annually among approximately 500 institutions (NCRR 1997). Anecdotal estimates indicate that the annual rate of increase in the use of rats and mice is about 20% (annual workshop of animal resources directors of major American universities, 1998), which suggest that annual use in the United States this year will exceed 15,000,000. Additionally, a recent poll by the Howard Hughes Medical Institute of approximately 60 constituent sites suggested that rodent use will increase by 1 to 3 times current levels during the next decade (J. Alford, Administrative Manager, Howard Hughes Medical Institute, personal communication during presentation at 1998 annual meeting of the Animal Resources Directors of Major American Universities).

Scientists treasure reproducibility. The word that describes what scientists do—research—implies the necessity of reproducibility for scientific investigation. The study of living things, a fundamental activity of biomedical scientists, presents a perennial challenge for attaining and sustaining reproducibility, especially when research involves complex organisms such as rodents. For the first two trimesters of the current century, genetic variability and rampant infection made animal research a risky business. However, steady progress in improving the health and genetic quality of laboratory mice and rats since the 1960s has reduced this risk and enhanced the value of animal experimentation in virtually every field of biology and medicine. Great progress occurred not only in the detection, elimination, and prevention of common pathogens, but also in the genetic manipulation of the mouse and rat through sophisticated breeding schemes. These advances provided access to novel mutants such as widely used models of immune dysfunction. Thus, as the century draws to a close, laboratory mice and rats have become vanguards of animal-based research. They are small enough (and big enough), tame enough, fertile enough, cheap enough, healthy enough, and genetically uniform enough to meet critical standards for mammalian modeling.

These attributes were less obvious for a while, not very long ago. Advances in molecular biology and biotechnology, especially during the 1970s and 1980s, were viewed by some as a harbinger of reduced reliance on vertebrate animal experimentation. In vitro or invertebrate alternatives offered opportunities for cheaper and faster answers to some scientific questions. In fact, many institutions experienced a decline in animal research during those decades. This trend occurred despite the fact that genetics, neoplasia, immunology, metabolism, and a host of other areas remained well suited to exploration in vertebrate models. But doubts about the relevance of vertebrate animal research are now moot, because, ironically, many of the tools and concepts that suggested imminent tempering of animal-based research became stimulants for an explosive growth in animal use. Molecular and developmental biologists put the mouse genome "in motion" and changed the face and potential of animal-based research dramatically and permanently. The impact of the genetically altered mouse, which is still a scientific infant, and its cousin, the genetically altered rat, which is by comparison a scientific fetus, is reflected in their anointment, with "digestible" hyberbole, as the *"E. colis"* of the 21st century.

The advent of genetically altered rodents, however promising scientifically, also is associated with biological, technological, logistical, and financial challenges that are emerging at an astounding rate. The challenges for assuring biologic integrity in genetically altered animals are dealing with intervening infections in diverse environments using diverse assessment standards and diverse terminology. Most of these challenges stem from the development, characterization, production, distribution, housing, husbandry, and health care associated with novel animals. And they raise a fundamental question: With so many genetically *new* animals being developed and used in so many places, by so many people, so quickly, how can their *biological integrity* be defined and ensured? The following remarks attempt to highlight briefly some of the issues flowing from this question with the expectation that others at this meeting will address them in greater depth.

My definition of "biological integrity" is incomplete, but, for the moment, consider the term to mean "the stability of intrinsic and extrinsic factors that define the structural and functional characteristics of an animal." Therefore, the benchmarks for defining a laboratory rodent in the era of genetic engineering must include at least the establishment, standardization, and monitoring of factors such as genotype, phenotype, microbial status, and environmental quality. Criteria such as reproductive capacity and other health-related factors such as susceptibility to infection should also be considered.

These concepts also imply that biological integrity can be perturbed by intrinsic or extrinsic interference, which may be overt or subtle. This threat is especially relevant considering the diversity of settings in which genetically engineered rodents are being made. Variability can be caused by genetic drift; the influence of genetic background on the penetrance of a phenotypic trait; opportunistic infection that may be pathogenic, disruptive to normal responses, or conducive to erroneous phenotyping; environmental stresses such as noise, vibration, and threatening odors; and many other factors. Variability also can be abetted by diverse or ill-defined terminology. For example, and as noted elsewhere in these proceedings (Lindsey, 1999), the term "specific pathogen free" has lost value because of the lack of precision with which it often is employed and perceived. Additionally, the increased use of animals inherently increases risks to biological integrity from dense housing and increasing exchanges of animals and animal products among laboratories, nationally and internationally.

Because worldwide reliance on laboratory rodents will increase for the foreseeable future, internationally standardized criteria and definitions should be developed as benchmarks for the biological integrity of laboratory rodents. A number of questions should be answered in formulating a transnational strategy to achieve this goal, a few of which are cited here. What are the criteria and definitions that should be used to measure biological integrity? Which assessments should be performed and how often? Who should perform the assessments? How should assessment results be reported and accessed? What sources are available to support research and development of new or improved assessment methods? Who should be responsible for funding assessment programs and how can the funds be leveraged for maximum benefit to biomedical research and the health of laboratory animals?

The time is ripe for international cooperation and action on these important issues. Meetings such as the US/Japan conference in session today can and should play a central role in getting planning under way.

REFERENCES

Animal resources directors of major American universities. 1998 annual workshop. Cincinnati, Ohio.
Lindsey, J. R. 1999. Current status of pathogen status in mice and rats. Pp. 39-43 in Proceedings of the US/Japan Meetings, October 23, 1998. National Academy Press, Washington, DC.
NCRR [National Center for Research Resources], National Institutes of Health. 1997. The national survey of laboratory animal use, facilities and resource. USPHS, Washington, DC.

Quality Testing System for SPF Animals in Japan and Problems in the Management of Such Systems

Toshio Itoh
Deputy Director, ICLAS Monitoring Center
Central Institute for Experimental Animals
Kawasaki, Japan

ROLE OF THE ICLAS MONITORING CENTER IN THE QUALITY CONTROL SYSTEM OF LABORATORY ANIMALS

The International Council for Laboratory Animal Science (ICLAS), the only international organization related to laboratory animal science, designated our institute as an ICLAS Monitoring Center in 1979. It is currently the only such center.

In Japan, most of the mice and rats used in experiments are specific pathogen-free (SPF) animals supplied by breeders. In animal experimentation facilities, barrier systems for maintenance of SPF animals are also widespread. However, there are no uniform standards in academic associations concerning quality testing systems for SPF animals. Several organizations have prepared recommendations for a quality testing system including test items, test frequency, and sample size. Those organizations are the ICLAS Monitoring Center, the Association of Laboratory Animal Facilities of National Universities, and the Japanese Society of Laboratory Animals. Breeders and users have established their own quality testing systems using these recommendations as a reference. However, because the staff of the ICLAS Monitoring Center participated in the preparation of these associations' recommendations, the quality testing systems of organizations that actively undertake quality control are basically the same as that of the ICLAS Monitoring Center.

The organization of the ICLAS Monitoring Center is as follows. An Advisory Board has been established in the Center to hear outside opinions. The members of this Board are the following six organizations: the Association of

Laboratory Animal Facilities of National Universities, Japanese Association for Experimental Animal Technologists, Japanese Association for Laboratory Animal Science, Japanese Association of Experimental Animals, Japanese Society of Laboratory Animals, and Japan Pharmaceutical Manufacturers Association. They include the main groups of laboratory animal breeders and users. The Center consists of three divisions: genetics, microbiology, and embryo bank. The operating funds are obtained as income from monitoring and cryopreservation services ordered by animal facilities of commercial breeders, pharmaceutical companies, universities and research institutes, as well as government support. Last year the Center received support from the Ministry of Education, Science, Sports and Culture of Japan.

In 1997, the activities of the microbiology division of the Center were as follows. Microbiological monitoring was performed on about 18,000 samples from about 1,700 animal production facilities and animal experimentation facilities. The Center also produced and supplied antigens, antisera, and antibody testing kits as reference substances indispensable in microbiological monitoring. About 1,000 vials of antigens and antisera and about 3,000 testing kits were distributed. The Center held two workshops with academic societies, gave lectures in universities and institutions, and jointly held training courses with various organizations in an effort to promote monitoring.

The antibody testing kit uses enzyme-linked immunosorbent assay (ELISA) produced and supplied by the Center and can be used for testing four microbes: Sendai virus, mouse hepatitis virus, *Mycoplasma pulmonis,* and Tyzzer's organism (*Clostridium piliforme*). These items were selected because of their importance as pathogens and their prevalence in Japan. The Center, which is the only organization in Japan that has its own antigens and antisera and system for microbiological testing, performs testing on a third party basis.

Overseas, the East Asian countries are still in a rather weak position, but their economies have expanded remarkably in recent years. In these countries, substantial progress has been made in science and technology. Assistance provided by the Center in the field of laboratory animals in these countries includes receiving trainees, on-site education, guidance, and a supply of reference substances. Since 1979, the Microbiology Division has accepted 16 trainees from Asian countries. By means of these activities in Japan and abroad, the ICLAS Monitoring Center has become a center for the quality control of laboratory animals, not only in Japan but also in East Asia.

MICROBIOLOGICAL MONITORING SYSTEM OF THE ICLAS MONITORING CENTER

There do not appear to be any major differences in sampling size and frequency of monitoring between Japan and the West, but there are slight differences in the criteria used for selection of test items. In the United States and especially Europe, all microorganisms that might affect experimental results are tested. However, we also take into consideration the pathogenicity of the microorganism for the animal, the possibility of transmission and disease in humans, and the opportunity for infection based on the contamination map. Because laboratory animals kept in barrier facilities are monitored, we do not think it is necessary to include parasites that require an intermediate host or microorganisms for which spontaneous infections have not been confirmed and the possibility of infectious disease has been found only in infection experiments. Monitoring also requires economic considerations. I do not think it is necessary to monitor all items at all times in all animal facilities; nor is it necessary for the tests to be the same for breeders and users. For example, breeders should supply as much information (test results) as possible on the animals of interest to the user since it is not clear for what experiments the animals will be used. However, researchers as users need only test results for microorganisms that might cause damage to the experiment. The quality control systems are basically different for animal experiments using immunodeficient animals and for those with little burden placed on the animals.

Our concept for selection of test items in the microbiological monitoring system of the Center is based on categorization of test items as described in the *Manual of Microbiologic Monitoring of Laboratory Animals* (US Public Health Service/NIH 1994). For the reasons I mentioned above, the microorganisms we test for are classified into five categories. Category A consists of zoonotic and human pathogens carried by animals; category B consists of pathogens fatal to animals; category C consists of pathogens that are not fatal but can cause disease in animals and affect their physiological functions; category D consists of opportunistic pathogens; and category E consists of indicators of the microbiological status of an animal or colony. The microorganisms to be tested should be selected based on the degree of microbiological control in each animal facility. For example, in SPF animal production facilities, as many items as possible including all categories are selected; in animal facilities requiring strict microbiological control, such as those performing experiments using immunodeficient animals or experiments placing a heavy burden on the animals, categories A, B, C, and D are selected; and in facilities undertaking experiments with little burden on the animals, categories A and B are sufficient. The test items performed in our Center on mice and rats are listed by category in Table 1. In the selection of test items in individual animal facilities, consideration should be given to the possibility of in-house testing, outsourcing of the testing, current status of microbiological contamination, and which experiments are being performed.

MICROBIOLOGICAL CONTAMINATION OF LABORATORY ANIMALS IN JAPAN

I present here our recent test results and compare them with those of the

TABLE 1 Selected Microbes for Monitoring

Category[a] Microbes (no. of organisms)		Mice 24 microbes	Rats 24 microbes
A(4)	Dermatophytes	X	X
	Hantavirus		X
	Lymphocytic choriomeningitis (LCM) virus	X	
	Salmonella spp.	X	X
B(5)	*E. coli* 9115a, c:k(B)	X	
	Ectromelia virus	X	
	Mouse hepatitis virus	X	
	Mycoplasma pulmonis	X	X
	Sendai virus	X	X
C(18)	*Clostridium piliforme*	X	X
	Bordetella bronchiseptica		X
	Cilia-associated respiratory (CAR) bacillus	X	X
	Corynebacterium kutscheri	X	X
	Giardia muris	X	X
	H-1 virus		X
	Kilham rat virus		X
	Minute virus of mice	X	X
	Mouse encephalomyelitis virus	X	X
	Mouse adenovirus	X	X
	Pasteurella pneumotropica	X	X
	Pneumonia virus of mice	X	X
	Reovirus type 3	X	X
	Sialodacryoadenitis virus		X
	Spironucleus muris	X	X
	Streptococcus pneumoniae		X
	Heliobactor hepaticus	X	
	Lactic dehydrogenase virus	X	
D(2)	*Pseudomonas aeruginosa*	X	X
	Staphylococcus aureus	X	X
E(1)	*Syphacia* spp.	X	X

[a]Category A: Pathogens that might infect humans.
Category B: Pathogens fatal to animals.
Category C: Pathogens not fatal, but can cause diseases in animals and affect their physiological functions.
Category D: Opportunistic pathogens.
Category E: Indicators of the microbiological status of an animal or colony.

United States. The facilities that asked the Center to perform the tests included breeders and animal experimentation facilities. Our results reflect the microbiological quality of laboratory animals in Japan.

There are three large and several small SPF animal breeders in Japan. The microbiological quality of animals in these SPF animal breeders has basically been maintained in good condition free from test items of categories A, B, C, and

E. However, infections do occur in SPF animal breeders. Contamination by *M. pulmonis* or *Pasteurella pneumotropica* has recently been seen in several SPF breeders.

Our results in mouse and rat experimental facilities can be seen in Tables 2 and 3. Among our categories A, B, and C, category A: zoonosis was never found, but contamination by pathogens in categories B and C have been observed in

TABLE 2 Microbiological Monitoring in Mouse Experimental Facilities (1992-1996)

		Facilities			
Category[a]	Items	Pharmaceutical companies[b]	%	Universities/ Institutes[b]	%
B	Mouse hepatitis virus	52/599	8.7	221/910	23.2
	Sendai virus	1/599	0.2	8/910	0.9
	Mycoplasma pulmonis	2/599	0.3	20/910	2.2
C	Pneumonia virus of mice	3/599	0.5	2/910	0.2
	Mouse encephalomyelitis virus	0/599		2/910	0.2
	Mouse adenovirus	0/599		1/910	0.1
	Clostridium piliforme	0/599		1/910	0.1
	Corynebacterium kutscheri	0/599		1/910	0.1
	Pasteurella pneumotropica	17/288	5.9	55/352	15.6
	Giardia muris	0/215		1/222	0.5
	Spironucleus muris	0/215		2/222	0.9
D	*Pseudomonas aeruginosa*	59/288	24.2	50/352	20.2
	Staphylococcus aureus	80/155	51.1	26/70	37.1
E	*Syphacia obvelata*	12/215	5.6	11/222	5.0

		Facilities
A	Dermatophytes	0/209
	Hantavirus	0/51
	Salmonella spp.	0/640
	Lymphocytic choriomeningitis (LCM) virus	0/79
B	Ectromelia virus	0/1509
	E. coli 0115a, c:k(B)	0/640
C	Minute virus of mice	0/62
	Cilia-associated respiratory (CAR) bacillus	0/64
	Helicobacter hepaticus	4/12

[a]Category A: Pathogens that might infect humans.
Category B: Pathogens fatal to animals.
Category C: Pathogens not fatal, but can cause diseases in animals and affect their physiological functions.
Category D: Opportunistic pathogens.
Category E: Indicators of the microbiological status of an animal or colony.

[b]No. of positive facilities/no. of tested facilities

TABLE 3 Microbiological Monitoring in Rat Experimental Facilities (1992-1996)

Category[a]	Items	Facilities Pharmaceutical companies[b]	%	Universities/ Institutes[b]	%
B	Sendai virus	3/694	0.3	14/315	4.4
	Mycoplasma pulmonis	1/694	0.1	29/315	9.2
C	Pneumonia virus of mice	2/694	0.3	1/315	0.3
	Sialodacryoadenitis virus	0/694		20/315	6.3
	Cilia-associated respiratory (CAR) bacillus	0/19		1/15	6.7
	Clostridium piliforme	38/694	5.5	28/315	8.9
	Corynebacterium kutscheri	2/694	0.3	0/315	
	Pasteurella pneumotropica	2/337	0.6	2/315	0.6
	Giardia muris	0/204		1/60	0.5
	Spironucleus muris	1/204	0.6	3/60	1.7
D	*Pseudomonas aeruginosa*	62/337	18.4	15/315	4.8
	Staphylococcus aureus	103/145	66.9	7/21	33.8
E	*Syphacia muris*	12/204	5.9	21/49	45.7

Category[a]	Items	Facilities
A	Dermatophytes	0/186
	Hantavirus	0/279
	Salmonella spp.	0/652
	Lymphocytic choriomeningitis (LCM) virus	0/26
C	H-1 virus	0/37
	Kilham rat virus	0/38
	Mouse adenovirus	0/1009
	Mouse encephalomyelitis virus	0/1009
	Bordetella bronchiseptica	0/337
	Streptococcus pneumoniae	0/652

[a]Category A: Pathogens that might infect humans.
Category B: Pathogens fatal to animals.
Category C: Pathogens not fatal, but can cause diseases in animals and affect their physiological functions.
Category D: Opportunistic pathogens.
Category E: Indicators of the microbiological status of an animal or colony.

[b]No. of positive facilities/no. of tested facilities

mouse experimental facilities. The highest contamination rates are seen for mouse hepatitis virus and *P. pneumotropica*. The number of positive items and their positivities are lower in pharmaceutical companies than in universities and institutes. In mouse experimental facilities, infections have been decreasing with the spread of the barrier system, and such infections have basically disappeared in pharmaceutical companies. However, in universities and research institutes where introduction of the barrier system has been delayed, there are still sporadic infections.

In rat experimental facilities, among our categories A, B, and C, category A was never found; but contamination by pathogens in categories B and C has been observed. Main pathogens detected in rats were *M. pulmonis*, *Clostridium piliforme*, cilia-associated respiratory (CAR) bacillus, and sialodacryoadenitis virus. The positive items and their positivities showed the same trends as those in mouse experimental facilities.

The microbiological quality of mice and rats between US and Japanese experimental facilities is shown in Figures 1 and 2. US results were quoted from "Health Care for Research Animals Is Essential and Affordable" in *FASEB Jour-*

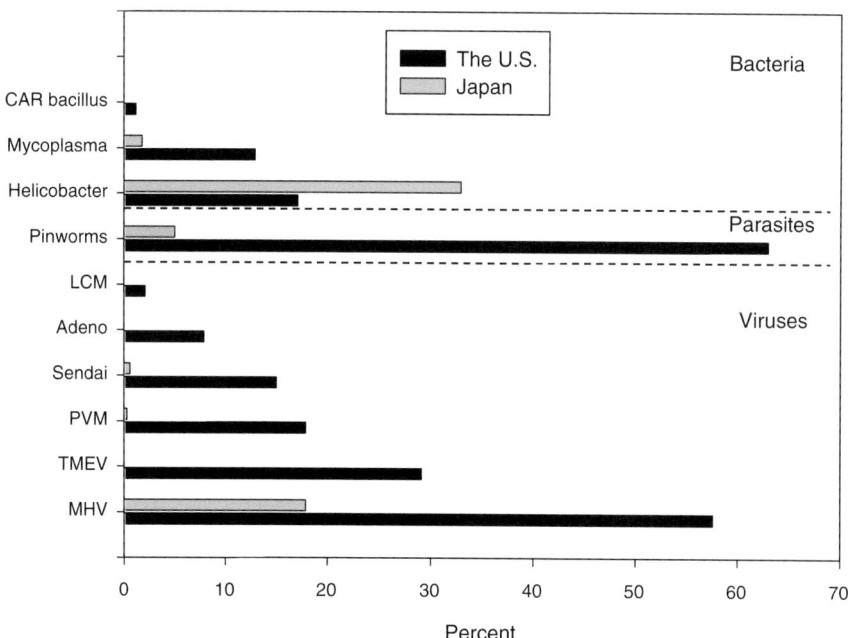

FIGURE 1 Comparison of microbiological quality of mice between the United States and Japan (% positive of agents in animal facilities).

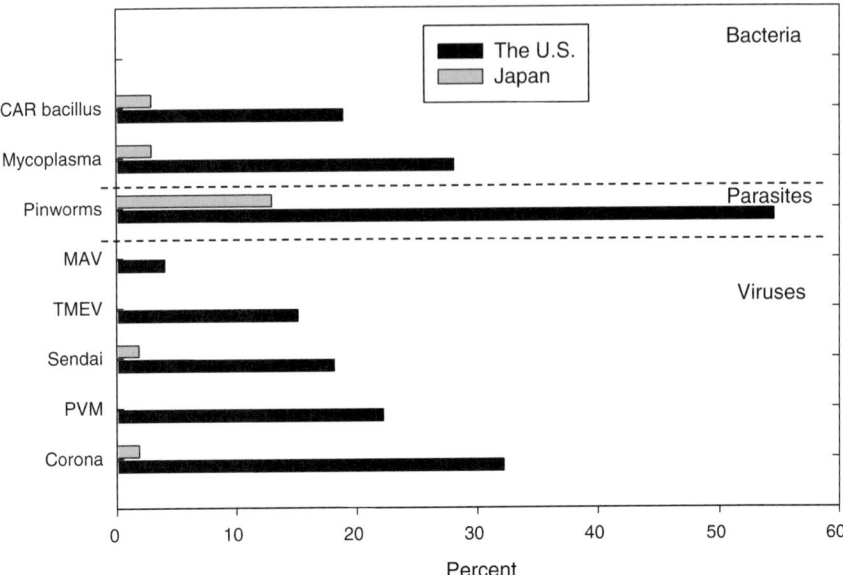

FIGURE 2 Comparison of microbiological quality of rats between the United States and Japan (% positive of agents in animal facilities).

nal by Drs. Jacoby and Lindsey (1997), who are present today. These results were compared for tests performed in both countries. The trends were the same for mice and rats. There were fewer positive items in Japan than in the United States, and when the tests were positive, the positive rates were lower in Japan.

Japan is a small country where it is comparatively easy to reach a consensus. Once such a consensus is reached, there is a tendency to persevere in the direction decided. This national characteristic may be the reason that quality control of laboratory animals in Japan has been more successful than in the United States.

CONCLUSION

Internationalization of laboratory animals has made remarkable advances as seen with genetically engineered animals, and mice and rats can be shipped all over the world by air. From the standpoint of animal control, there are now more opportunities for infected animals or materials obtained from infected animals to enter facilities. In Japan currently, *Pneumocystis carinii* pneumonia and viral hepatitis in immunodeficient mice and in immunological function knockout mice introduced from the United States have become a major problem in microbiological control in animal facilities. In the United States, contamination of sera with ectromelia virus has presented a problem. Animal quality control, especially microbiological monitoring, is becoming more important for maintenance of laboratory animals and assuring reproducibility of experimental results.

REFERENCES

Jacoby, R., and J. R. Lindsey. 1997. Health care for research animals is essential and affordable. FASEB J. 11:609-614.

US Public Health Service/NIH. 1994. Manual of Microbiologic Monitoring of Laboratory Animals. 2nd edition. (NIH Publication No. 94-2498). GPO, Washington, D.C.

Definition of Microbiological Status of Rats and Mice / The Need for Methods of Defining Flora / International Standards for Terminology

Kazuaki Mannen
Associate Professor, Laboratory Animal Research Center
Oita Medical University
Japan

CURRENT STATUS OF MICROBIOLOGICAL QUALITY OF LABORATORY ANIMALS IN UNIVERSITY ANIMAL CENTERS IN JAPAN

The organization of university animal centers in Japan is shown in Figure 1. The national university animal centers consist of 53 facilities. The Division of Science and International Affairs Bureau of the Ministry of Education, Science, Sports, and Culture, of which Mr. Ota is Director, is closely linked with all of the animal centers.

In Japan, the national university medical schools and national institutions are members of the Association of Laboratory Animal Facilities of the National University in Japan, hereafter referred to as University Facilities Japan (UFJ). We occasionally encounter subtle differences between the required microbiological inspection of animals being transported among the national, public, and private colleges and other, atypical locations such as other academic institutions or nonapproved vendors and researchers. This type of problem also exists with international transportation. Because specific pathogen-free (SPF) animals are under strict microbiological control, it is technically not necessary to consider pathogenic contamination during their transportation to and from UFJ locations. However, we have found during quarantine inspection that gene-manipulated animals, such as transgenic and knockout mice and rats, have been contaminated by some microorganisms. For this reason, UFJ has established a Working Biohazard Committee (of which I am a member) to formulate guidelines for the microbiological quarantine inspection of mice and rats.

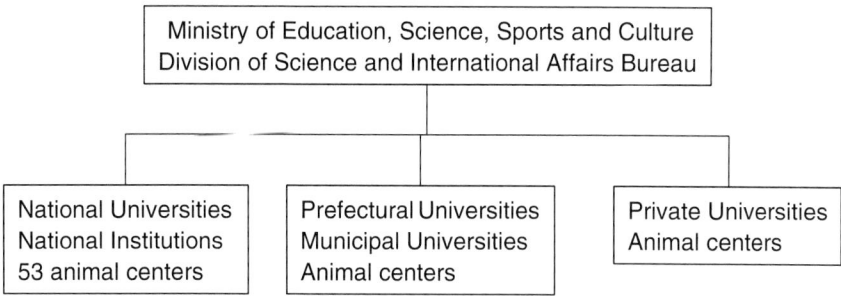

FIGURE 1 Organizational chart of animal centers in Japanese universities.

IMPORTANCE OF IDENTIFYING CONTAMINATION

Forty-nine of the 53 facilities at national universities and centers participated in the survey (Table 1). Of the 49 facilities, 47 (96%) had gene-manipulated mice from domestic sources. A total of 28 facilities (57%) had mice from international sources, mainly from the United States and in some cases from Great Britain, Switzerland, France, Canada, and Germany.

The contaminating microorganisms of the mice are shown in Table 2. Major organisms were mouse hepatitis virus, *Pasteurella*, *Mycoplasma*, *Syphacia*,

TABLE 1 Introduction of Transgenic and Knockout Mice (1996-1997)

Number of facilities responding: 49 (n=53)
Facilities with transgenic and knockout mice: 47 (96%)
    Domestic introduction: 47 (96%)
    International introduction: 28 (57%)

U.S.A.
American Red Cross Holland Laboratory
Charles River Lab
Chrysalis DNX Transgenic Sciences
Harvard University
Jackson Lab
NCI
NIH
McLaughlin Research Institute
Northwestern University
North Carolina University
Stanford University
University of Missouri College of Vet Med
University of California

Great Britain
Mammalian Genetic Unit
Medical Research Council

Switzerland
CIBA

France
Institut Gustave Roussy

Canada
Ontario Cancer Institute

Germany
Heidelberg Universitat

TABLE 2 Contamination at the Introduction of Transgenic and Knockout Mice (1995-1997)

Domestic Introduction	International Introduction
Mouse hepatitis virus (11)	*Pasteurella* spp (12)
Pasteurella pneumotropica (10)	*Trichomonas* spp (9)
Mycoplasma pulmonis (5)	*Pneumocystis carini* (8)
Syphacia spp (5)	Mouse hepatitis virus (5)
Pseudomonas aeruginosa (4)	Duodenum & Cecum for protozoa (4)
Sendai virus (3)	*Helicobacter* spp (4)
Staphylococcus aureus (2)	Mouse poliovirus (GDVII) (3)
Trichomonas spp (2)	*Proteus* spp (3)
Aspiculuris tetraptera (1)	*Actinobacillus* spp (2)
Bordetella branchiseptica (1)	Mouse rotavirus (EDIMV) (2)
Corynebacterium kutscheri (1)	*Klebsiella* spp (2)
Hanta virus (HFRS) (1)	Mouse parvovirus (2)
Octomitus pulcher (1)	*Pseudomonas aeruginosa* (2)
Pneumocystis carinii (1)	*Staphylococcus aureus* (2)
Polyplax spinulosa (1)	*Syphacia obvelata* (2)
Salmonella spp (1)	Cecal amebiasis (1)
Streptococcus zooepidemics (1)	*Entamoeba muris* (1)
Myobia musculi (1)	*Myobia musculi* (1)
	Myocoptes musculinus (1)
	Streptococcus spp—hemolytic (1)
	Theiler's virus (1)
	Carbacillus (1)

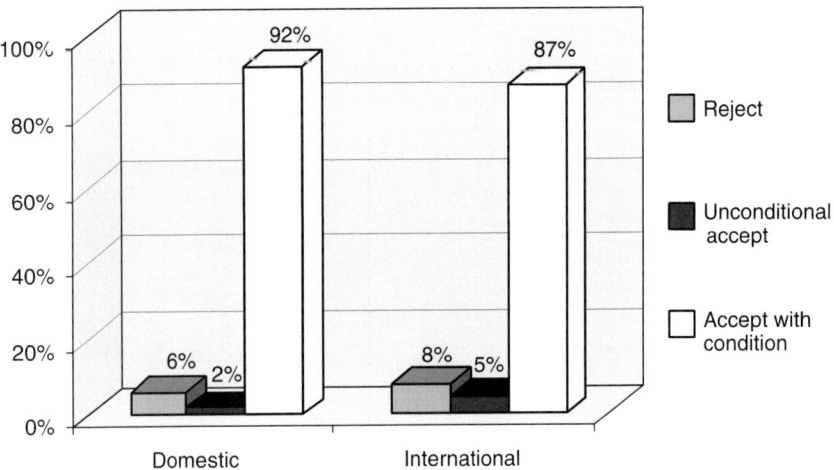

FIGURE 2 Contamination at the introduction of TG/KO mice between 1995 and 1997.

Pseudomonas, Sendai virus, and so forth from domestic introduction. In contrast, the spectrum of contaminating organisms resulting from international transportation differed from domestic transportation.

As shown in Figure 2, almost all (87 to 92%) of the contaminated mice were accepted conditionally. (Surprisingly, 2% of domestic and 5% of internationally transported mice that were contaminated were accepted unconditionally.) Of the 137 domestic cases, 17% were cleaned up after acceptance; 19% were confined in special rooms; and 63% were not specified. Of the 103 international cases, 61% were confined in special rooms—roughly three times more than in domestically contaminated mice. Although the reason for this difference is unclear, it may reflect the facility administrators' belief that international contamination is a more serious problem.

Development of Rodent Pathogen Profiles and Adequacy of Detection Technology

Steven H. Weisbroth
President, AnMed/Biosafe, Inc.
Rockville, Maryland

Dr. Ralph Dell, attendees of the US/Japan Meeting, and readers, as well, will have interest in being reminded that Dr. Howard Schneider, then Chairman of ILAR, in his opening remarks to an international symposium said that "Those familiar with the scientific process will recognize at once that this is only temporary, that one can confidently predict further progress by the time of the Fifth International Symposium." The symposium he was addressing was the Fourth, subtitled "Defining the Laboratory Animal," given here in this city in 1969, and eventuating in a text of the proceedings under the same name (ICLA 1971). Thus, 30 years later, here we are discussing the same themes under the same subtitle. However, since then, a great deal of progress has been made, as predicted by Dr. Schneider. I paraphrase my colleague Dr. David Baker who, in a recent review (Baker 1998), characterized progress in control of infectious disease in laboratory rodents as follows:

> Around the turn of the century, an investigator might have said: "I can't do my experiment today, all of my rats are dead." In the 1960's he might have said: "I can't do my experiment today, all of my rats are sick"; and in the 1990's he might have said: "I can't do my experiment today, all of my rats are antibody positive."

A cynic might predict that in the year 2000 an investigator may have to say, "I can't do my experiment today, my IACUC won't let me use rats." (This last statement, of course, is not attributable to Dr. Baker.)

From the beginning, more than 100 years ago, until the present, issues of rodent health have remained as important concerns, not only to investigators, but also to the service personnel charged with the production, procurement, and care of these animals. In attempting to put dimensions on the diversity of laboratory animal disease, it is first necessary to understand that each animal species is host to an etiologic spectrum composed of arthropod ectoparasites; helminth and protozoan endoparasites; and fungal, bacterial, rickettsial, and viral forms more or less associated by common experience and as documented in the literature as indigenous to that host species. What the history of laboratory rodents seems to demonstrate is that this spectrum as a concept is not a list frozen for all time, but rather more closely represents a moving boundary in which old pathogens are eradicated, creating invasive opportunities for new pathogens and, thus, periodic reconstitution of the lists (Weisbroth 1996). In practice, the process moves gradually and the cast of characters is adjusted as circumstances and diagnostic experience warrant. Over the years, the principal effects of disease control, eradication, and exclusion programs have been to reduce both the range of diversity and the incidence of agents listed in the panels. There appears little doubt that at present we are witness to a major restructuring of the list of indigenous pathogens of laboratory rodents, a process accelerated by the highly structured and microbially limited environments permitted by production environments and good laboratory animal practice.

For the purpose of defining the microbial health status of laboratory rodents and lagomorphs, comprehensive health surveillance programs are oriented to the systematic diagnostic examination of sample groups of animals against a predetermined list of pathogenic organisms. The pathogens are organized into etiologic classes to form panels of the more common (or classically associated) indigenous agents. Collectively, these panels form the microbial definition of the status high quality research rodents are expected to meet. Sample groups statistically representative of the larger group from which they are drawn and meant to define are tested for the presence, or absence, of the specific agents making up the lists or panels. Properly conducted, findings in the sample groups can be used to infer presence of the detected agents in the larger population they represent.

For purposes of monitoring the status of closed breeding colonies and of resident populations at user institutions, a program oriented to scheduled repetitive testing on an ongoing basis is developed. It is difficult to overstate the importance of scheduled, repetitive testing that not only provides current, timely information about the health status of specific subpopulations, but also because results on the sample groups form an additive sample size over time, increasing confidence that negative results in the sample groups truly are representative of the population as a whole. This concept is particularly important in assessing the reliability or accuracy of negative results. The strategy of health surveillance testing is oriented to detection of even a single positive instance, since such a finding implies that the larger population has been likewise exposed and may be contaminated with that agent. Conversely, failure to detect, or negative results for an agent, form the objective basis on which to conclude that the unit has not been exposed to that agent. Scheduled repetitive testing increases the sample size drawn from the unit and strengthens confidence that continued negative results are not a product of sampling error or test inadequacy.

It is also important to recognize that the agents vary in prevalence; thus for closed units, it is unnecessary and uneconomical to uncritically and evenly apply the same testing frequency to all agents in the maximal panels. These considerations, so-called "smart testing," lead health monitoring program managers to use reduced or "core" panels to monitor the more prevalent agents on a more frequent basis and maximal comprehensive panels on a less frequent basis, depending on the needs of the particular program.

I would like to turn our attention now to a consideration of the agents making up testing panels themselves and adequacy of testing methodology to detect these agents, at this point in time. I hasten to explain that while the panels presented here have no official standing, they would with minor exceptions represent a consensus of expert opinion in the United States.

The genera of important rodent louse and mite ectoparasites are listed in Table 1. Direct low power microscopic examination (10×) of cadavers shortly after euthanasia, at the level of the base of the hairs and skin, has proven to be the most rapid and accurate means of assessing ectoparasitic status. A more time-consuming, but technically acceptable, alternative is to allow the cadavers to cool while on black paper, allowing motile forms to come out to the surface or crawl off onto the paper where they may be picked up by cellophane tape, placed on slides, and identified under the microscope. Examination of skin scrapings is not

TABLE 1 Arthropod Ectoparasites—Diagnostic Alternatives for Rodent Comprehensive Health Surveillance Profiles

Ectoparasite	Standard	Acceptable Alternate or Adjunctive	Not Recommended
Myobia	1	2	3
Myocoptes	1	2	3
Radfordia	1	2	3
Psorgates	1, 3		
Notoedres	1	2	3
Demodex	3		4
Liponyssus	1	2	3
Polyplax	1	2	3

[1.] Direct visualization of skin and pelage by low power microscopy
[2.] Motile forms on hair tips and on black paper under cooling carcass
[3.] Skin scraping
[4.] Skin section

a reliable method to establish ectoparasitic status. In general, for the arthropod parasites, present test methodology is adequate for surveillance programs.

Table 2 is a listing of helminth endoparasites. The pinworms *Syphacia* and *Aspicularis* may be diagnosed alternatively by direct examination of the dissected cecum and colon by low power (3 to 30×) microscopy, or by fecal flotation, or both. Fecal flotation has the advantage of allowing simultaneous detection of (*Eimeria*) coccidia but could miss an early, preovulatory helminth infestation. As with the ectoparasites, present methodology is satisfactory for surveillance programs.

The protozoa of importance are listed in Table 3. Readers familiar with the Federation of European Laboratory Animal Associations (FELASA) agent panels (Kraft and others 1994; Rehbinder and others 1996) may note the absence of *Klossiella* from this list, which has been deleted because it has not been reported from US colonies in modern times. There will be more to say about the FELASA listings as I continue. With the exception of the coccidia, the other enteric protozoa on this list are motile flagellates easily detected in temporary wet mounts of intestinal scrapings by microscopic examination (100×) or in histologic sections of small intestine. The coccidia, however, require fecal flotation for accurate detection of low-level infection. The hemoprotozoa require blood films for detection of parasitized cells but may be inapparent in latently infected immunocompetent hosts. The hemoprotozoa have not been reported in this country in many years, like lactic dehydrogenase elevating virus (LDHV), probably because of essential eradication of their (required) hematophagous arthropod vectors. *Encephalitozoon* infection is easily detected by present enzyme-linked immunosorbent assay assay (ELISA) serology. The latter test is sufficiently reliable to form the basis for "test and removal" eradication programs. Histopathology for *Encephalitozoon* detection is a useful confirmatory adjunct, but

TABLE 2 Helminth Endoparasites—Diagnostic Alternatives for Rodent Comprehensive Health Surveillance Profiles

Helminth	Standard	Acceptable Alternate or Adjunctive	Not Recommended
Aspicularis	1	3	4
Syphacia	1	3	4
Hymenolepis	3		4
Trichosomoides	2	4	

[1.] Direct visualization of lumen contents of cecum and colon by low power microscopy
[2.] Direct visualization of lumen surface of urocyst by low power microscopy
[3.] Fecal flotation in hypertonic solution
[4.] Microscopic examination of histologic sections of relevant tissues

TABLE 3 Protozoa—Diagnostic Alternatives for Rodent Comprehensive Health Surveillance Profiles

Protozoa	Standard	Acceptable Alternate or Adjunctive	Not Recommended
A. Enteric Forms			
Giardia	1	3	
Spironucleus (Hexamita)	1	3	
Entamoeba	1	3	
Trichomonas, Tritrichomona	1	3	
Eimeria	4		1, 3
B. Hemoprotozoa			
Hemobartonella	2		
Eperythrozoon	2		
C. Other			
Encephalitozoon	5	3	

1. Microscopic examination of wet mounts of intestinal scrapings
2. Microscopic examination of stained blood films
3. Microscopic examination of histologic sections of relevant tissues
4. Fecal flotation in hypertonic solution
5. Serologic immunoassay

many infected individuals do not develop chronic lesions. The indicated methodology for detection should be regarded as adequate for the protozoa and sufficient to support surveillance programs.

Table 4 is a panel of classical bacterial pathogens of laboratory rats and mice. The indicated "standard" methods for detection should be regarded as adequate for surveillance programs in terms of accuracy of detection. Nonetheless, as time goes on, the polymerase chain reaction (PCR) is being more frequently applied for more faster and economical means of detection. The approach of using PCR for genome detection is particularly apt because the conventional methodology likewise requires detection of the microbes themselves, rather than immuno-serologic indicators of exposure (antibodies), which does form the primary detection mode for the viruses. Note that serology is used, however, as the standard test mode of screening programs for rodent *Mycoplasma* infections and the cilia-associated respiratory bacillus for which serology forms a more useful screening device than cultural isolation. *Bordetella* was listed as a nod to tradition; certainly it has not been a reported pathogen of laboratory rats and mice for many years, if ever.

The most common and important viral pathogens of laboratory mice and rats are listed in Tables 5 and 6, respectively. It will be seen that with the exception

TABLE 4 Bacteria And Mycoplasmas—Diagnostic Alternatives for Rodent Comprehensive Health Surveillance Profiles

Bacteria and Mycoplasma	Standard	Acceptable Alternate or Adjunctive
Salmonella	1, 2	
Streptobacillus moniliformis	1	
Streptococcus pneumoniae	1	
Streptococcus, B-hemolytic	1, 2	
Cilia-associated resp. bacillus	4	3
Mycoplasma pulmonis	4	1, 3, 5, 6
Mycoplasma arthritidis	4	1, 3, 5, 6
Bordetella bronchiseptica	1, 2	
Pasteurella pneumotropica	1, 2	5
Pseudomonas aeruginosa	1, 2	5
Citrobacter rodentium	1, 2	5
Klebsiella pneumoniae	1, 2	5
Klebsiella oxytoca	1, 2	5
Staphylococcus aureus	1, 2	5
Corynebacterium bovis (HAC)	1, 2	3, 5

1. Broth and agar media for primary isolation and semi-differentiation
2. Microtized media strips for biochemical profile and identification
3. Microscopic examination of histologic sections of relevant tissues
4. Serologic immunoassay
5. Polymerase chain reaction (PCR)
6. Corroborating gross pathology

of Riley's LDHV, testing strategy for the presence of these viruses is oriented to detection of antibodies engendered by viral infection. The assumption is (validly) made that viral infections reflect themselves by antibody production, and the presence or absence of the virus in the colony may be directly inferred by the presence or absence, respectively, of antibodies in the sample groups. With several exceptions, discussed below under Problematic Issues, the ELISA and immunofluorescent assay (IFA) are adequate methodology to support surveillance programs for all of the indicated agents (Lussier and others 1991). Indeed, the adequacy of these tests for surveillance programs has permitted essential eradication of all of these agents from quality breeding stocks in the United States.

There are, however, diagnostic situations in which the actual presence or absence of the murine virus in test samples must be determined. These situations have classically been investigated by isolation in tissue culture or by the mouse (or rat) antibody production (MAP or RAP) test (Lussier and others 1991). Increasingly, because of the minimum 3 to 4 week time required to conduct a

TABLE 5 Mouse Viruses—Diagnostic Alternatives for Rodent Comprehensive Health Surveillance Profiles

Viruses		Test Modes[a]		
		Std	Alt	Other
PVM	Pneumonia Virus of Mice	E	I	
REO3	Respiratory Enteric Orphan III	E	I, H	
SEN	Sendai Virus	E	I	
GD7	Theiler's Murine Encephalomyelitis Virus (TMV)	E	I	
LCMV	Lymphocytic Choriomeningitis	E	I	
HAN	Hantaan Virus	E	I	
MVM	Minute Virus of Mice	I	E	H, PCR
MPV	Mouse Parvovirus	I	E	H, PCR
MHV	Mouse Hepatitis Virus	E	I	PCR
KV	Kilham's Virus	E	H	
EDIM	Epidemic Diarrhea of Infant Mice	E	I	
MAV	Mouse Adenovirus	E	I	
ECTR	Ectromelia Virus	E	I	
POLY	Polyoma Virus	E	I	
MCMV	Mouse Cytomegalovirus	E	I	
MTV	Mouse Thymic Virus	I		
LDHV	Lactic Dehydrogenase Elevating Virus	C		

[a]Std = Standard or Preferred, Alt = Alternate or Confirmatory, E = Enzyme Linked Immunosorbent Assay (ELISA), C = Biochemical Assay, I = Immunofluorescent Assay (IFA), H = Hemagglutination Inhibition (HAI)

TABLE 6 Rat Viruses—Diagnostic Alternatives for Rodent Comprehensive Health Surveillance Profiles

Viruses		Test Modes[a]		
		Std	Alt	Other
PVM	Pneumonia Virus of Mice	E	I	
REO3	Respiratory Enteric Orphan III	E	I	
SEN	Sendai Virus	E	I	
GD7	Theiler's Encephalomyelitis or TMI	E	I	
LCMV	Lymphocytic Choriomeningitis	E	I	
HAN	Hantaan Virus	E	I	
KRV	Kilham's Rat Virus	E	I	H
RPV	Rat Parvovirus	E	I	H
TH1	Toolan's H1 Virus	E	I	
SADV/RCV	Sialodacryoadenitis Virus/Rat Corona Virus	E	I	

[a]Std = Standard or Preferred, Alt = Alternate or Confirmatory, E = Enzyme Linked Immunosorbent Assay (ELISA). I = Immunofluorescent Assay (IFA), H = Hemagglutination Inhibition (HAI)

MAP test, PCR applications are being employed for detection of viral contaminants of biotechnical products and tissue culture cell lines, and for investigation of disease outbreaks. PCR applications presently have their greatest utility in testing for a single virus, and a growing body of literature attests to the use of PCR for murine virus detection. However, the versatility of the MAP test simultaneously permits detection of all 16 to 18 agents of concern. Research is currently under way to enable genomic detection of viral groups or multiplex tests to deal with this problem.

Finally, it needs to be accepted that there is a residual group of agents not likely to be detected by traditional screening tests of nonlesioned rodent sample or sentinel groups. These agents, along with the *Helicobacter* species, share the feature of being difficult or impossible to isolate using routine microbiologic methods or to histopathologically demonstrate in the absence of lesions. They can be described as latent and clinically silent, and this term is used to list them in Table 7. Traditional means of detection for this important group have required first rendering the carrier hosts immunodeficient by chemical immunosuppressants (such as cortisone [and its synthetic derivatives]) or antimetabolites (such as cylosphosphamide) so as to encourage clinical recrudescence or florid expression of the pathogens, if present. This approach—the stress test—has been used particularly to reliably demonstrate closed rodent production units as free of the Tyzzer's disease agent (*Clostridium piliforme*), *Corynebacterium kutscheri,* and *Pneumocystis carinii* (Weisbroth, 1995). It is with this group that PCR has had its greatest utility in rodent disease detection. Properly conducted, the stress test is arduous, takes 3 to 4 weeks to complete, and is expensive. By comparison, PCR can be done in 1 to 2 days, does not require ablation of the normative immune inhibition of pathogen populations, and is at least 1 to 2 logs more

TABLE 7 Latent and Clinically Silent Agents Probably not Detected by Standard Methodology—Diagnostic Alternatives for Rodent Comprehensive Health Surveillance Profiles

Agents	Preferred	Acceptable Alternate or Adjunctive	Not Recommended
Clostridium piliforme	1	2	3
Corynebacterium kutscheri	1	2, 3	
Pneumocystis carinii	1	2	
Helicobacter sp.	1	4, 3	

[1] Polymerase chain reaction (PCR)
[2] Stress test with terminal samples processed by standard microbiologic and histopathologic methods
[3] Serologic immunoassay
[4] Specialized microbiologic isolation methods

sensitive. PCR should be regarded as the preferred surveillance methodology for this group, with properly conducted stress tests also being considered as methodologically adequate.

PROBLEMATIC ISSUES

I have indicated serology as "not recommended" for the diagnosis of *Clostridium piliforme* for several reasons. On the one hand, there is no basis in the literature for the use of serology as a screening device to detect nonlesioned carriers. More important, on the other hand, is the seeming predilection of rodents (and rabbits) to carry natural antibodies, ostensibly to commensal *Clostridia*, which induce low-level cross-reactive positivity to *C. piliforme* antigens in both ELISA and immunofluorescent assay test modes. Uncritical acceptance of this type of positivity has led to interpretation of animal colony contamination with *C. piliforme*, when in fact this is not the case. Any provisional diagnosis of *C. piliforme* infection on the basis of serology alone needs supportive corroboration by PCR, or histopathology, or both to confirm the diagnosis, or it is likely to be wrong.

At present, a similar situation complicates serodiagnosis with a number of the murine viruses, in which the rodent host serum is tested with antigen to a murine virus, and antibodies to a different and presumptively human origin virus cause mainly low-titered cross-reactive positivity. Such is the situation with antibodies to REO III (REO3), Sendai (SEN), and SV5 in guinea pig serum from certain colonies, with reactivity to the GDVII (GD7) strain of Theiler's encephalomyelitis virus (TMEV) in certain rat sera and, in the author's opinion, antibodies to the agents now termed rat and mouse parvovirus, respectively, in rat and mouse serum. Thus, at present, the problematic issues with murine virus serology are not sensitivity, as they were with the earlier generation of complement fixation and hemagglutination inhibition methods, but rather, issues of specificity. Gradually it is becoming apparent that the problems of rodent virus infection are rather more complicated than simply being limited to indigenous rodent virus infection. These cross-reacting and confusing serologic reactions are nature's way of telling us that these rodents are being exposed to other viruses, often under circumstances in which a contaminated host source other than the humans who come in contact with the animals is difficult to credibly posit. We can expect this issue (that is, the human-to-rodent interface) to remain a complicating issue in rodent diagnostics and rodent health until the animal care community comes to grips with regulating the interface better than we do at present.

In concluding my remarks, I would like to address a number of points that relate to a global perspective in rodent health assessment.

1. Can we have a globally universal standard, in terms of panels of agents of which the rodents must be free to be accepted as the highest quality? In other words, could we truly have global harmonization of an infectious standard for each of the laboratory rodents? The answer is that between countries and regions, there would of course be substantial areas of overlap as there now is between the lists presented here as representing a US consensus and the published official FELASA lists. But equally true, there would probably always be discrepant agents of regional concern on the panels not regarded as significant elsewhere. As examples, the FELASA lists for rats and mice include *Klossiella*, *Proteus*, *Leptospira*, *Escherichia coli,* and *Yersinia pseudotuberculosis*—which you will note are not in the US panels presented here—whereas *Pasteurella pneumotropica* and *Citrobacter rodentium* are considered of importance in the United States but not by FELASA. Perhaps expert committees could be formed to critically examine whether discrepant agents need to retained or could be safely deleted from the regional panels.

2. There is at present no means for ensuring a uniform standard for the potency, purity, or specificity of serologic test reagents, either of antigens or of positive control sera. Similarly, there is at present no means for ensuring availability of testing reagents for diagnostic labs. There should be, perhaps at the ICLAS level, some objective means of comparing and evaluating adequacy and availability of testing reagents to remove this potential variable to comparability of surveillance programs from laboratory to laboratory, and from country to country.

3. The testing laboratories themselves, whether at the state, national, or private sector level, are not regulated and required to meet administered performance standards. Whether required for professional acceptability (dare I say accredited?) or by entirely voluntary participation, there should be some objective ongoing assessment of laboratory performance.

4. Finally, I would like to give an opinion against patent protection for discoveries of new agents. An unfortunate trend has been the movement to cash in on diagnostic discoveries by patenting organisms as isolated from nature and subsequently characterized as pathogens. An example is *Helicobacter hepaticus*. We have seen that the patent on this agent has acted to restrict its availability for exploration and implementation of diagnostic tests to the detriment of improvements in rodent health surveillance. Presumably this was not the intention of the patent holders, but just as surely, that has been one of the net effects. The unintended effect of restricted availability should be noted and the impulse to patent such "discoveries" discouraged.

REFERENCES

Baker, D.G. 1998. Natural pathogens of laboratory mice, rats and rabbits and their effects on research. Clin. Microbiol. Rev. 11:231-266.

ICLA [International Committee on Laboratory Animals]. 1971. Defining the Laboratory Animal. IV Symposium. National Academy of Sciences, Washington, DC.

Kraft, V., A. A. Deeny, H. M. Blanchet, R. Boot, A. K. Hansen, G. Milite, J. R. Needham, W. Nicklas, A. Terrot, C. Rehbinder, Y. Richard, and G. De Vroey. 1994. Recommendations for the health monitoring of mouse, rat, hamster, guinea pig and rabbit breeding colonies. Lab. Anim. 28:1-12.

Lussier, G. L., J. K. Davis, W. R. Shek, A. L. Smith, and G. Lussier. 1991. Detection methods for the identification of rodent viral and mycoplasmal infections. Lab. Anim. Sci. 41:199-225.

Rehbinder, C., P. Baneux, D. Forbes, H. Van Herck, W. Nicklas, Z. Rugnya, and G. Winkler. 1996. FELASA recommendations for the health monitoring of mouse, rat, hamster, gerbil, guinea pig and rabbit experimental units. Lab. Anim. 30:193-208.

Weisbroth, S. H. 1995. *Pneumocystis carinii*: Review of diagnostic issues in laboratory rodents. Lab. Anim. 24:36-40.

Weisbroth, S. H. 1996. Post-indigenous disease: Changing concepts of disease in laboratory rodents. Lab. Anim. 25:25-33.

Current Status of Pathogen Status in Mice and Rats

J. Russell Lindsey
Professor, Department of Comparative Medicine
University of Alabama Schools of Medicine and Dentistry
Birmingham, Alabama

Following the pattern set by the previous Japanese speakers, Drs. Itoh and Mannen, Dr. Weisbroth has addressed the issue of "testing methods," and I will speak on "current status" of pathogens in mice and rats. I am, of course, most familiar with the quality of animals in my own country, the United States. However, many of my comments will be applicable worldwide as biomedical research and its constituent parts, including the laboratory animals being used to generate much of the data and the journals that disseminate the results, are increasingly international activities.

PAST PROGRESS TOWARD REDUCING PATHOGENS

Much progress has been made in eliminating pathogens from laboratory mouse and rat populations since the 1960s (Weisbroth 1996). This fact is borne out by several surveys conducted mainly in the early 1980s (NRC 1991, pp. 7-8) and one in 1988 (Casebolt and others 1988). During the period from the 1960s into the 1990s, there were also major efforts to (1) improve methods for detecting, eliminating, and preventing pathogen infections (Allen and Nomura 1986; Baker 1998; Bhatt and others 1986; NRC 1991); and (2) herald the seemingly endless detrimental effects that pathogens have on research results (Allen and Nomura 1986; Bhatt and others 1986; NRC 1991 [see partial listings on pp. 274-275]). More recently, many additional examples of pathogen effects have been published (Baker 1998). Although all of these past developments are noteworthy achievements that tend to engender a great sense of accomplishment among laboratory animal specialists, it seems to me far more important to take the forward view and ask the question, "How well are these achievements serving contemporary research?" The answers to this question are rather disappointing for a number of reasons.

PATHOGENS STILL POSE PERVASIVE RISKS IN THE UNITED STATES

Dr. Jacoby and I recently surveyed the top 100 institutional recipients of National Institutes of Health (NIH) funds regarding their efforts to prevent pathogen infections in mouse and rat populations at their institutions. Responses were received from 72 of the 100 institutions, representing more than $5 billion in NIH support and use of 3 million mice and 1 million rats for the year of the survey (Jacoby and Lindsey 1997, 1998). The results were surprising. Only 70% of the mouse populations and 60% of the rat populations were considered specific pathogen free (SPF), meaning that many populations were, in reality, conventional. Furthermore, the survey was constructed so that each institution could define its use of the term SPF by checking presence or absence of each of 24 pathogens. The results showed enormous disparities among institutions in the definition of SPF. Agents such as ectoparasites, parvoviruses, mouse hepatitis virus, mouse rotavirus, *Helicobacter* sp., Theiler's encephalomyelitis virus, and cilia-associated respiratory bacillus, generally recognized as pathogens, were common in so-called SPF populations. There was also great variation in the frequency of testing for pathogens, the number of animals tested, and the test methods being used. These findings point to real problems with the terminology of pathogen status.

PATHOGEN STATUS GETS LOST IN THE TERMINOLOGY MORASS

Unfortunately, the hard reality is that pathogen status is elusive because it is usually defined by words selected from a morass ("a marsh, swamp, or something that traps, confuses, or impedes") (*Webster's Ninth New Collegiate Dictionary* 1989) of terminology, including words such as germ free, defined flora, pathogen free, specific pathogen free, virus antibody free, barrier maintained, and conventional. The underlying problem is that these terms merely identify concepts. The results of the terminology morass are (1) there is no universal testing strategy or reporting terminology for clear and consistent definition of pathogen status in rodent populations, and (2) each institution (or investigator) selects its own list of pathogens, test procedures, animal sampling strategy, frequency of sampling, and reporting terminology. Definitions of the conceptual terms above should include the agents for which tests were actually done on the subpopulation(s) in question, test methods used, results of the tests, and frequency of the testing.

PRINCIPLES OF PATHOGEN STATUS ARE BEING COMPROMISED

Based on the foregoing information, it appears that a number of key principles of rodent pathogen status are being compromised:

1. Infectious agents that have been documented to cause significant disease or to interfere with research results need to be recognized as pathogens in the context of research use.
2. Pathogens of rodents are notorious for causing subclinical infections.
3. Many subclinical pathogen infections cause altered research results.
4. No meaningful inferences can be made about pathogen status without the benefit of results from tests for specific agents.
5. Immunodeficient animals require different test batteries than immunocompetent animals.
6. Tests for pathogens are performed on subpopulations of animals, not on "facilities."

Terms such as "pathogen free" are actually concepts and must be defined each time they are used (Allen and Nomura 1986; Bhatt and others, 1986; NRC 1991).

Some of the reasons these principles are being compromised in the United States are as follows:

1. The NIH National Center for Research Resources (NCRR) supported diagnostic and investigational laboratories were phased out (NRC 1998).
2. The Office of Management and Budget cost accounting standards shifted payment for diagnostics to investigators (NRC 1998).
3. Regulatory issues are requiring increasing attention (NRC 1998).
4. Rodent populations are increasing dramatically (transgenics, immunodeficients) (Jacoby and Lindsey 1997, 1998).
5. Transfers of animals between laboratories are increasing worldwide (Jacoby and Lindsey 1997, 1998).
6. Overt clinical disease is not widespread (all rodent pathogens tend to cause subclinical infections).
7. Investigators are becoming increasingly focused on molecular events and less aware of integrative biology.
8. There is little appreciation of pathogen importance among investigators, as evidenced by the following.

SCIENTISTS HAVE LITTLE APPRECIATION OF PATHOGEN STATUS

Since Dr. Jacoby and I published the results of the survey of institutions in the United States on their programs for protecting mouse and rat populations

TABLE 1 Documentation of Pathogen Status in Scientific Articles

Journal (volume)	Status defined by test results	PF,[a] SPF,[a] etc., used, not defined	No status information	Total papers reviewed
Inf. Immun.[a] (65)	3	23	74	100
J. Immunol.[a] (158)	4	18	78	100
Am. J. Pathol.[a] (150-151)	3	16	81	100
P.N.A.S.[a] (94)	1	4	95	100
Science (275-280)	0	0	100	100

[a]Am. J. Pathol., *American Journal of Pathology*; Inf. Immun., *Infection and Immunity*; J. Immunol., *Journal of Immunology*; PF, pathogen free; P.N.A.S., *Proceedings of the National Academy of Sciences U. S. A.*; SPF, specific pathogen free.

from pathogens (Jacoby and Lindsey 1997, 1998), I have often wondered what affect rodent health testing or the lack thereof has on the quality of the results appearing in scientific journals, and how investigators and journal editors perceive the importance of pathogen status. As a result, I chose five journals and reviewed in each of them the first 100 original scientific articles using mice or rats, beginning with their January 1997 issue. I searched each article for information defining pathogen status (Table 1). The journals were *Infection and Immunity*, *Journal of Immunology*, *American Journal of Pathology*, *Proceedings of the National Academy of Sciences U. S. A.*, and *Science*. The results clearly indicate that investigators and journal editors have little appreciation of pathogen status.

In summary, let me simply reiterate some of the main "take home" messages about the current status of pathogen status of mice and rats used in research. Although progress has been made over the past 40 years in reducing their prevalence, rodent pathogens still present serious risks to a substantial amount of the research data being generated every year. Unfortunately, there are many obstacles to further reductions in pathogen prevalence, including problems of terminology, adherence to sound principles of pathogen exclusion, lack of financial support for health surveillance testing, and lack of appreciation by investigators and journal editors of the deleterious effects pathogens have on research results. Overall, the data point to the conclusion that progress toward elimination of pathogens is not keeping pace with the current sophistication in biomedical science.

REFERENCES

Allen, A. M., and T. Nomura, eds. 1986. Manual of Microbiologic Monitoring of Laboratory Animals. NIH Pub. No. 86-2498 (USDHHS, NIH, DRR). Government Printing Office, Washington, D.C.

Baker, D. G. 1998. Natural pathogens of laboratory mice, rats, and rabbits and their effects on research. Clin. Microbiol. Rev. 11:231-266.

Bhatt, P. N., R. O. Jacoby, H. C. Morse III, and A. E. New, eds. 1986. Viral and Mycoplasmal Infections of Laboratory Rodents: Effects on Biomedical Research. Academic Press; Orlando, Fl. 844 pp.

Casebolt, D. B., J. R. Lindsey, and G. H. Cassell. 1988. Prevalence rates of infectious agents among commercial breeding populations of rats and mice. Lab. Anim. Sci. 38:327-329.

Jacoby, R. O., and J. R. Lindsey. 1997. Health care for research animals is essential and affordable. FASEB J. 11:609-614.

Jacoby, R. O., and J. R. Lindsey. 1998. Risks of infection among laboratory rats and mice at major biomedical research institutions. ILAR J. 39:266-271.

NRC [National Research Council]. 1991. Infectious Diseases of Mice and Rats. A report of the Committee on Infectious Diseases of Mice and Rats. National Academy Press, Washington, D.C. 397 pp.

NRC [National Research Council]. 1998. Biomedical Models and Resources: Current Needs and Future Opportunities. A report of the Committee on New and Emerging Models in Biomedical and Behavioral Research. National Academy Press, Washington, D.C. 45 pp.

Webster's Ninth New Collegiate Dictionary. 1989. Merriam-Webster Inc., Springfield, MA. P. 771.

Weisbroth, S. H. 1996. Post-indigenous disease: Changing concepts of disease in laboratory rodents. Lab. Anim. 25:25-33.

Genetic Background and Phenotypes in Animal Models of Human Diseases

Kazuo Moriwaki
Vice President, The Graduate University for Advanced Studies
Kanagawa-ken, Japan

DEVELOPMENT OF EXPERIMENTAL MOUSE STRAINS

In the field of mammalian genetics including human genetics, the effect of genetic background on the expression of a particular gene for a given biological function or disease has long been a well-known but unresolved subject. More than two decades ago, Goodenough and Levine (Goodenough and Levine 1974) foresaw that a particular gene product would normally operate in the presence of countless different combinations of other gene products. Because we did not have a dependable method of mapping multiple genes, considerable effort was invested in developing experimental strains with the same genetic background as the chromosomal region to be analyzed. The most valuable contribution was the establishment of H2 congenic mouse strains (Snell and others 1976), in which the structure and immunological function of mouse major histocompatibility complex (MHC) could be clearly demonstrated. These findings have also resulted in very useful models of human MHC, particularly the structures and functions relevant to human diseases.

As a result of recent remarkable developments in the technology of both gene and embryo manipulation, we can now isolate genes for a biological function or human or mouse diseases as DNA molecules and inject them into the mouse in an early embryo stage to observe their expression in the whole body. The successful establishment of embryonic stem (ES) cell lines has also made it possible to knock out a given gene in the embryo that later develops to adulthood.

TRANSGENIC AND KNOCKOUT MICE

The development of modern technology has shed light on the rather classical problem of genetic background. Of the large number of reports that have been published recently on transgenic and knockout experiments in mice, many have described significant effects of genetic background (that is, strain specificity). Threadgill and his colleagues (Threadgill and others 1995) demonstrated the effect of strain difference on the embryonic lethality in the EGFR gene-targeted mouse. The CF-1 strain with the targeted gene died at a much earlier stage that the CD-1 strain. Sibilia and Wagner (Sibilia and Wagner 1995) showed the strain-dependent epithelial defects in mice lacking epidermal growth factor receptor (EGFR). Those mice with 129/Sv genetic background died at the mid-gestation stage, whereas those with 129/Sv × B/6 × MF/1 survived to postnatal 20 days. Wolf and Henderson (Wolf and Henderson 1998) recently reported the effect of strain difference in the transgenic introduction of the human P450 gene in mouse, which can be expressed in the C3H strain but not in the BALB strain.

RECOMBINANT INBRED STRAINS

Recombinant inbred (RI) strains have been developed for mapping of a specific gene that has different alleles between the two parental strains based on the strain distribution pattern. Bailey (Bailey 1971) first conceived the usefulness of RI strains for analyzing multiple genes controlling biological functions and diseases. When he established CXB RI strains, however, he learned that the number of marker gene loci was not enough to map one or more genes precisely. Many RI strains have been developed recently, and they can be used for mapping multiple gene loci by use of microsatellite DNA primers, the polymerase chain reaction technique, and computer software for quantitative trait locus [QTL] analysis. Although these technical advances have also made it possible to map multiple gene loci by conventional backcrosses, more accurate mapping (and complete homozygosity in their recessive alleles) can be done by employing RI strains, as discussed by Silver (Silver 1995).

Nishimura and colleagues (Nishimura and others 1995) have established the new 21 SMXA RI strains from SM/J and A/J progenitor strains. By using those RI strains, Pataer and colleagues (Pataer and others 1997) recently identified a new gene locus for the resistance to urethan-induced pulmonary adenomas. Susceptibility to the pulmonary adenoma has so far been considered to be controlled by at least four genes (Festing and others 1994): (1) Pas1 linked to Kras2 on number 6 chromosome, (2) Pas2 to MHC on number 17, (3) Pas3 to D9Mit11 on number 9, and (4) Pas4 to D19Mit16 on number 19. Moreover, two dominant resistant genes, Par 1 on number 11 and Par2 on number 18, have been reported (Manenti and others 1996; Obata and others 1996).

COMMON DISEASE MODELS

Development of the modern mapping techniques described above has also made it possible to map multiple genes causing common adulthood model diseases in mice (for example, diabetes in non-obese diabetes [NOD] strain). From those studies, it is assumed that although most mutations have mild effects, a specific combination of them can facilitate the expression of an ethnological mutation. Because common adulthood (life-style) diseases such as diabetes and cancer appear to be caused by the specific combination of many normal variant genes and, in many cases, etiological genes, the animal models for them should replicate human disease states. A broader study of gene loci related to diseases requires more variant alleles in mice for analyzing the molecular mechanism of gene manifestation. Asian mice are useful for that purpose because they are genetically more remote from laboratory mice and have plenty of variant alleles. We were able to conduct a DNA analysis using 60 marker DNA loci with Asian mice (Moriwaki and others 1999).

The finding that variant genes contained in the Asian wild mice sometimes have a long evolutionary history is biologically important to investigate the mechanism of gene function. It is not possible to select for long evolutionary history in fancy mice and laboratory mice.

As seen in the NOD experiment conducted by Wakana and colleagues (Wakana and others 1997), genetic introduction of a genetically remote allele of Idd-4 in Asian wild-derived MSM strain (established from wild mice collected in Mishima) exhibited a significant increase in frequency of diabetes. This strain should be a useful model to analyze Idd-4 function, which cannot be observed by the introduction of BALB/c or C57BL/6 alleles.

Another example of the characteristic function of Asian wild-derived alleles is the expression of the Rim4 mutant phenotype, polydactyly, which was completely suppressed in the Asian wild-derived genetic background (Masuya and others 1997). One might expect some "dominant negative" structural change in the gene product.

CONCLUSION

Animal models of common adulthood diseases such as diabetes and cancer have indicated that these diseases are apparently caused by the specific combination of many normal variant genes and possibly some etiological genes. To further our knowledge requires additional animal models so that we can identify a large number of variant alleles that vary within the normal range. For this purpose, Asian wild-derived genes are useful not only for the number of variations, but also for the large differences in the genome structure, which sometimes give rise to a "dominant negative" effect. These characteristics are useful for analyzing the mechanism of normal gene functions as seen in the case of Rim4 mouse.

REFERENCES

Bailey, D. W. 1971. Recombinant-inbred strains. Transplantation 11:325-327.
Festing, M. F., A. Yang, and A. M. Malkinson. 1994. At least four genes and sex are associated with susceptibility to urethane-induced pulmonary adenoma in mice. Genet. Res. 64:99-106.
Goodenough, U., and R. P. Levine. 1974. Genetics. Holt, Rinehart and Winston, Inc., New York.
Manenti, G., M. Galibordi, R. Elango, A. Fiorino, L. De-Gregorio, F. S. Falvella, K. Hunter, D. Housman, M. A. Pierotti, and T. A. Dragani. 1996. Genetic mapping of a pulmonary adenoma resistance (Par 1) in mouse. Nat. Genet. 12:455-457.
Masuya, H. T. Sagai, K. Moriwaki, and T. Shiroishi. 1997. Multigenic control of the localization of the zone of polarizing activity in limb morphogenesis in the mouse. Dev. Biol. 182:42-51.
Moriwaki, K., N. Miyashita, Y. Yamaguchi, and T. Shiroishi. 1999. Multiple genes governing biological functions in the genetic backgrounds of laboratory mice and Asian wild mice. Prog. Exp. Tumor Res. Karger, Basel. 30:1-12.
Nishimura, M., N. Hirayama, T. Serikawa, K. Kanehira, Y. Matsushima, H. Katoh, S. Wakana, A. Kojima, and H. Hiai. 1995. The SMZA: A new set of recombinant inbred strain of mice consisting of 26 substrains and their genetic profile. Mamm. Genome 6:850-857.
Obata, M., H. Nishimori, K. Ogawa, and G. H. Lee. 1996. Identification of the Par2 (pulmonary adenoma resistance) locus on mouse chromosome 18, a major genetic determinant for lung carcinogen resistance in BALB/cByJ mice. Oncogene 13:1599-1604.
Pataer, A., M. Nishimura, T. Kamoto, K. Ichioka, M. Sato, and H. Hiai. 1997. Genetic resistance to urethane-induced pulmonary adenomas in SMXA recombinant inbred mouse strains. Cancer Res. 57:2904-2908.
Sibilia, M., and E. F. Wagner. 1995. Strain-dependent epithelial defects in mice lacking the EGF receptor. Science 269:234-238.
Silver, L. M. 1995. Mouse Genetics. Oxford University Press, New York.
Snell, G., J. Dausset, and S. Nathenson. 1976. Histocompatibility. Academic Press, New York.
Threadgill, D. W., A. A. Dlugosz, L. A. Hansen, T. Tennenbaum, U. Lichti, D. Yee, C. LaMantia, T. Mourton, K. Herrup, R. C. Harris, J. A. Barnard, S. H. Yuspa, R. J. Coffey, and T. Magnuson. 1995. Targeted disruption of mouse EGF receptor: Effect of genetic background on mutant phenotype. Science 269:230-234.
Wakana, S., T. Shiroishi, K. Moriwaki, A. Kono, and T. Nomura. 1997. Susceptibility gene Idd4 controls onset of IDDM: An allele from the nondiabetic MSM strain is associated with early onset of diabetes in mice. 11th Annual Mouse Genome Conference, Miami, Florida (Abstract).
Wolf, C. R., and C. J. Henderson. 1998. Use of transgenic animals in understanding molecular mechanisms of toxicity. J. Pharm. Pharmacol. 50:567-574.

Genetic and Phenotypic Definition of Laboratory Mice and Rats / What Constitutes an Acceptable Genetic-Phenotypic Definition

Hideki Katoh
Chief, Genetics Division
ICLAS Monitoring Center
Central Institute for Experimental Animals
Kanagawa, Japan

GENETIC STUDIES ON CLOSED COLONIES OF THE RAT

The following questions are important to all of us who are interested in human racial differences:

- Are the ancestors of present races common or different?
- How do we understand genetic relations among the races?

Population genetics studies provide answers to these questions as follows.

HUMAN ANCESTRY

We know that all human races have one common ancestor, whose progeny migrated across the world roughly 50,000 to 200,000 years ago. After these common ancestors settled in various areas and adapted to the environments, they developed as races.

That today's races have a common ancestor means that they had common genes. These races exist as a result of accumulated genetic changes caused by gene mutations and of gene frequencies at many loci. Such genetic changes resulted in racial diversity, with different genetic characteristics such as skin color and height.

Genetic Relations Among Races

Population geneticists have calculated gene frequencies of blood types and have shown similarities among races using differences of various blood type frequencies from blood group data of all of the races. Genes and their frequencies of common ancestors of the human races were randomly changed and varied when the ancestors dispersed to various parts of the world and settled there. Randomly occurring changes of gene frequency are called genetic drift. One example of genetic drift obtained by computer simulation can be seen in Figure 1. If the ancestors had A and B blood types at rates of 50% each, one of the two races will show a higher percentage of the A type and the other loses the A type after 300 generations. These changes are natural and occur by chance. Assuming a breeding cycle of 30 years, 300 generations is equivalent to about 10,000 years in humans. This time is sufficient to develop races different from the original one. However, it is important to remember that mating between races is possible, and such offspring will also be reproductive. This ability to produce fertile offspring means that no race has developed into subspecies. In the mouse, 1,000,000 years are required for genetic divergence to lead to a subspecies.

Ethnic Differences

It is interesting to consider the number of loci required to identify ethnic differences of human races. In Figure 2, the results for several racial groups can be seen using 12 loci, with Caucasoids, Mongoloids, Negroids, and Australoids clearly identified. These results agree with those of anthropology and cultural anthropology. What happens if we have fewer loci for an ethnic analysis? Figure 3

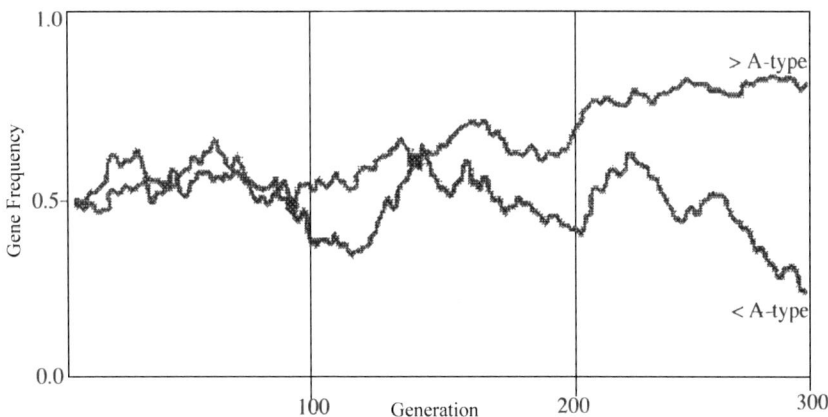

FIGURE 1 A computer-simulated result of genetic drift.

50 MICROBIAL AND PHENOTYPIC DEFINITION OF RATS AND MICE

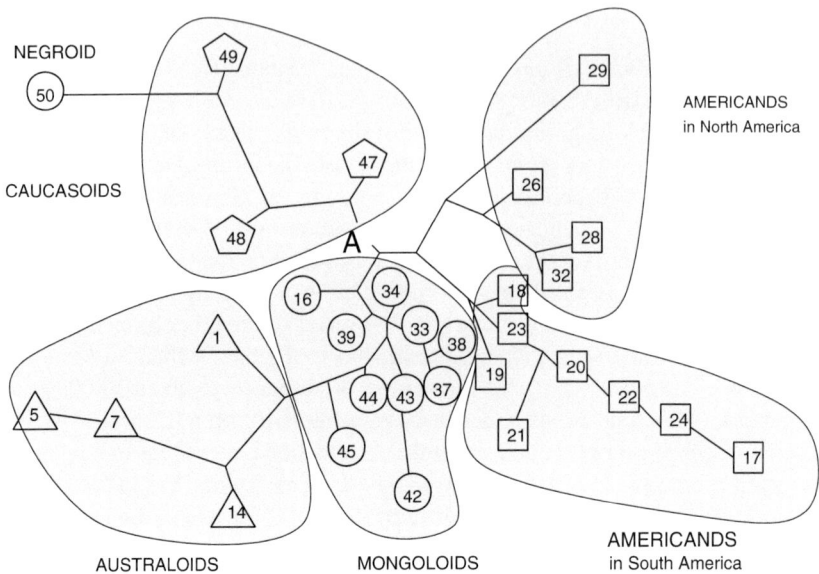

FIGURE 2 Genetic relations among 31 races obtained from the gene frequencies of 12 loci.

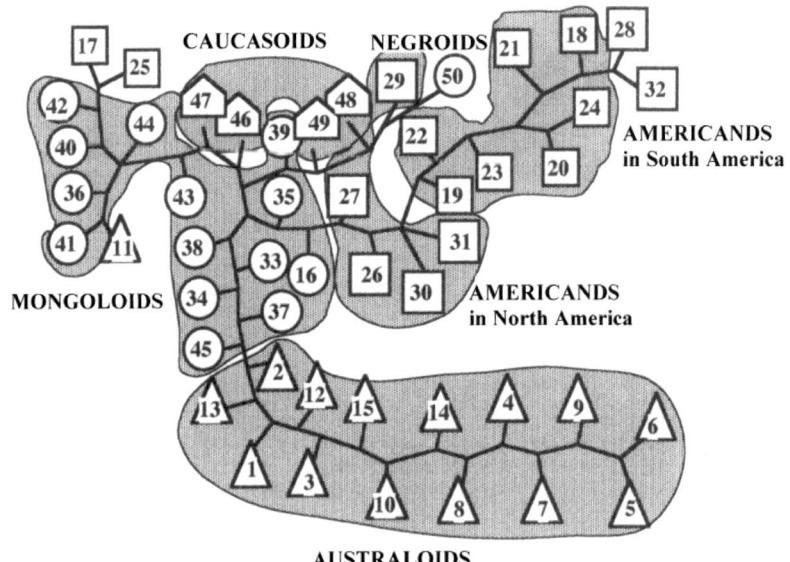

FIGURE 3 Genetic relations among 50 races obtained from gene frequencies of ABO, Rh, and MN blood types.

illustrates the results for the same racial groups using only three loci. As expected, the results are incomplete because some races are grouped into the wrong groups. Thus more reliable results are obtained in a population genetics study by using as many loci as possible.

LABORATORY ANIMALS

Closed Colonies

To apply the genetic information described above to laboratory animals, it is necessary to define "laboratory animals." Mice and rats include both laboratory and wild animals. Wild animals have been domesticated, and inbred strains and closed colonies have been bred as laboratory animals. In addition, there are many inbred strains and closed colonies with particular characteristics. As with humans, laboratory and wild animals also have common ancestors and so have inherited the same set of genes a long time ago.

At the beginning of the 20th century, inbreeding was started to obtain inbred lines of mice. The oldest lines in the mouse have approximately an 80-year history. Because mice and rats reproduce three times (generations) a year, an 80-year history is equivalent to 240 generations. We can anticipate that two lines derived from a single ancestor differ genetically in the same way as two human races differ. It is important to remember that in spite of the two lines being separated 80 years ago, they can still mate and produce fertile offspring.

However, in the case of closed colonies, mice and rats have approximately a 70-year history (Lindsey 1979; Morse 1981). Closed colonies also have several sublines derived from the main colonies. Genetic analysis reveals fewer differences among closed colonies of mice and rats than among humans, indicating less genetic drift.

One important problem exists in the case of laboratory animals, as illustrated in the following extreme example. Suppose that a colony of breeder A was supplied to breeder B 10 years ago. Genetic drift is sufficiently slow that we can assume that in 10 years, the two lines will not diverge dramatically. In fact, the two breeders do anticipate such changes. However, it is possible that breeding schemes can be accidentally mixed up by people working in animal facilities. It is possible that they will take males and females that are closely related, thereby establishing a subcolony in a breeding facility. In such a case, the two colonies will be genetically quite different, and a typical bottleneck effect will occur.

Consider a more ideal situation in a rat colony. Suppose that a researcher is working on allotransplantation of a cancer cell line using a closed colony. Assume that the cancer has the A-antigen of RT1. In the closed colony as a recipient of the cancer, the percentages of the animals with A, AB, and B were 25%, 50%, and 25% in one generation. It can be assumed that the cancer will be transplantable in 75% of the animals with the A-antigen on average. If animals with the

B-type contributed to the subsequent generation of the divided colony, they will produce only offspring carrying B, which will reject the cancer. Such cases generally do not arise, but the worse case (such as artificially lowering the frequency of A-antigen-positive animals) should be carefully considered.

DEMONSTRATION OF THE EXISTENCE OF CLOSED COLONIES USING GENETIC MONITORING

Closed colonies of rats are known to show genetic polymorphisms at many loci. When they are maintained under unplanned mating, the stocks may show genetic instability. One cause of instability is a bottleneck effect caused by caesarian section, followed by an increase in an inbreeding coefficient in the colony. In this case, the number of loci showing genetic polymorphisms decreases with each generation.

Generally, genetic variation in a closed colony is greater than that seen with an inbred strain. When a colony has been genetically altered for any of the reasons described above, it will be difficult for users of the stocks to interpret their data because genetic changes have occurred. Therefore, we should periodically test closed colonies using genetic methods to confirm genetic stability.

From the viewpoint of population genetics, we studied closed colonies of rats for 3 years (1988 to 1990). The Ministry of Education, Science, Culture, and Sports of Japan provided a grant for our study in which we performed genetic analysis on closed colonies of rats with the following aims: (1) to obtain genetic profiles of closed colonies of rats, (2) to obtain monitoring profiles of closed colonies, and (3) to determine specific genes identifying each colony. As mentioned above, an additional goal of this study was to establish methods of genetic monitoring that are appropriate for closed colonies.

Materials and Methods

As shown in Figure 4, six closed colonies were obtained from four breeders in Japan: Jcl:Wistar, Crj:Wistar, Iar:Wistar, Jcl:SD, Crj:SD, and Nr:Donryu. Sixty rats consisting of 30 females and 30 males of each colony were used for genetic study. Of the genetic markers used in the study, 21 are biochemical markers and one is an immunological marker (Table 1).

Results

Gene frequencies and specific markers identifying colony. Gene frequencies of 21 biochemical markers are shown in Table 2. Results for RT1 are summarized in Table 3. Markers showing > 0.017 genotypic frequency (equivalent to at least one occurrence in 60 animals tested) were counted. Markers identifying colonies were demonstrated as follows: $Acon1^a$: SD stocks; $Ahd2^b$:

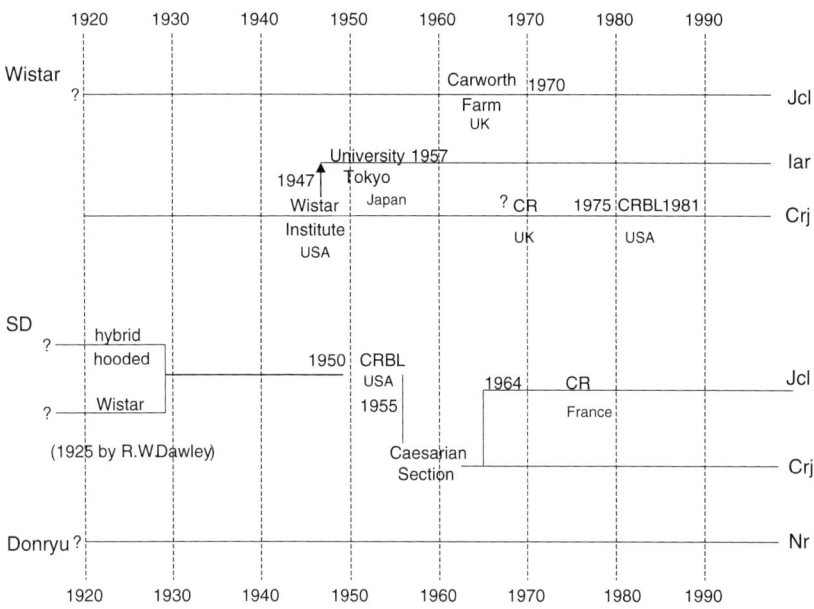

FIGURE 4. History of the five outbred stocks of the rat used in this study.

Jcl:Wistar; $Amy1^b$: Crj:SD; $Es3^b$: Iar:Wistar; $Es4^c$: Nr:Donryu; $RT1.A^k$: Iar:Wistar; $RT1.E^d$: Crj:Wistar.

Average heterozygosity. As shown in Table 4, percentages of polymorphic loci in six colonies varied from 9.5% (Jcl:Wistar) to 61.9% (SD), and average heterozygosities of six colonies ranged from 0.007 (Donryu) to 0.251 (Crj:SD). Because average heterozygosities of mice and human are 0.078 and 0.099, respectively (Nei 1987), those of rat closed colonies were reasonable. Jcl:Wistar revealed the lowest percentage (9.5%) of polymorphic loci. The breeder stated that this colony was reconstituted from several pairs by cesarian section to produce a SPF colony, which might have led to elevation of homozygosity at most loci. Donryu was second in percentage of polymorphic loci and first for average heterozygosity. This ranking was caused by a significant imbalance of allele frequencies of polymorphic loci in the Donryu colony.

Genetic distance. In Table 5 are shown the genetic distances among the colonies. Each value represents the level of genetic difference between two colonies. The lowest value was 0.151 between Jcl:SD and Crj:SD, and the highest one was 0.474 between Jcl:SD and Donryu. The close relation between Iar:Wistar and Donryu is shown by a distance value of 0.160.

TABLE 1 Marker Loci, Samples and Testing Methods

Gene Symbol		Gene Name	Samples	Testing Methods
Biochemical markers				
1	Acon1	Aconitase1	Kidney	CAE
2	Ahd2	Aldehyde dehydrogenase 2	Liver	IEF (pH3.5-10)
3	Ahdc	Aldehyde dehydrogenase c	Liver	IEF (pH3.5-10)
4	Akp1	Alkaline phosphatase 1	Kidney	IEF (pH5-7)
5	Alp1	Serum alkaline phosphatase 1	Kidney	IEF (pH5-7)
6	Amy1	Amylase 1	Pancreas	CAE
7	Es1	Esterase 1	Plasma	CAE
8	Es2	Esterase 2	Plasma	PAGE (10% gel)
9	Es3	Esterase 3	Small intestine	PAGE (10% gel)
10	Es4	Esterase 4	Kidney	CAE
11	Es6	Esterase 6	Testis	IEF (pH5-7)
12	Es7	Esterase 7	Testis	IEF (pH5-7)
13	Es8	Esterase 8	Testis	IEF (pH5-7)
14	Es9	Esterase 9	Testis	IEF (pH5-7)
15	Es10	Esterase 10	Testis	IEF (pH5-7)
16	Es14	Sex-influenced esterase	Plasma	CAE
17	Fh1	Fumarate hydratase 1	Kidney	CAE
18	Gc	Group specific component	Plasma	PAGE (10% gel)
19	Hbb	Hemoglobin beta chain	Red blood cells	CAE
20	Mup1	Major urinary protein 1	Urine	PAGE (15% gel)
21	Svp1	Seminal vesicle protein 1	Seminal vesicle fluid	CAE
Immunological marker				
	RT1	Histocompatibility 1	Red blood cells	Hemagglutination

CAE: Cellulose acetate membrane electrophoresis
PAGE: Polyacrylamide gel electrophoresis
IEF: Isoelectric focusing

SUMMARY AND DISCUSSION

The results of this study can be summarized as follows:

1. Genes uniquely identifying each colony were found.
2. Lower polymorphism observed in Jcl:Wistar was caused by caesarian section rederivation of the colony.
3. Genetic distance suggested that Crj:SD and Jcl:SD are the closest genetically. It was also revealed that the relation between Iar:Wistar and Donryu was very close yet genetically different from SD and Wistar colonies.

Although geneticists and breeders recognize the importance of genetic monitoring of closed colonies, this issue has not been discussed. The major reason is

TABLE 2 Gene Frequencies of 21 Biochemical Markers in Six Outbred Stocks

	Locus	Allele	Jcl:Wistar	Crj:Wistar	Iar:Wistar	Jcl:SD	Crj:SD	Donryu
1	Acon1	a	0.0000	0.0000	0.0000	**0.9580***	**0.5670***	0.0000
		b	1.0000	1.0000	1.0000	0.0420	0.4330	1.0000
2	Ahd2	b	**0.1830***	0.0000	0.0000	0.0000	0.0000	0.0000
		c	0.8170	1.0000	1.0000	1.0000	1.0000	1.0000
3	Ahdc	a	0.0000	1.0000	1.0000	1.0000	0.9020	1.0000
		b	1.0000	0.0000	0.0000	0.0000	0.0980	0.0000
4	Akp1	a	1.0000	1.0000	1.0000	1.0000	1.0000	1.0000
		b	0.0000	0.0000	0.0000	0.0000	0.0000	0.0000
5	Alp1	a	0.0000	0.0000	0.0000	0.0000	0.0000	0.0000
		b	1.0000	1.0000	1.0000	1.0000	1.0000	1.0000
6	Amy1	a	1.0000	1.0000	1.0000	1.0000	0.6500	1.0000
		b	0.0000	0.0000	0.0000	0.0000	**0.3500***	0.0000
7	Es1	a	0.0000	0.0000	0.7000	0.1920	0.3750	1.0000
		b	1.0000	0.9670	0.3000	0.0000	0.1670	0.0000
		c	0.0000	0.0330	0.0000	0.8080	0.4580	0.0000
8	Es2	a	1.0000	0.9170	0.0000	0.7670	0.3080	0.0000
		b	0.0000	0.0000	0.0000	0.0000	0.0000	0.0000
		c	0.0000	0.0000	0.5420	0.1670	0.0000	0.5920
		d	0.0000	0.0830	0.4580	0.0670	0.6920	0.4080
9	Es3	a	0.3330	0.9170	0.0000	0.2920	0.3920	0.9500
		b	0.0000	0.0000	0.4250	0.0000	0.0000	0.0000
		c	0.0000	0.0000	0.2330	0.4670	0.0000	0.0000
		d	0.6670	0.0830	0.3420	0.2420	0.6080	0.0500
10	Es4	a	0.0000	0.0000	0.5420	0.1580	0.0000	0.0000
		b	1.0000	1.0000	0.4580	0.8420	1.0000	0.1420
		c	0.0000	0.0000	0.0000	0.0000	0.0000	**0.8580***
11	Es6	a	1.0000	0.5520	1.0000	0.4330	0.6670	1.0000
		b	0.0000	0.4480	0.0000	0.5670	0.3330	0.0000
12	Es7	a	0.0000	0.0000	0.0000	0.0000	0.0000	0.0000
		b	1.0000	1.0000	1.0000	1.0000	1.0000	1.0000
13	Es8	a	0.0000	0.9170	0.4330	0.2330	0.7170	0.1330
		b	1.0000	0.0830	0.5670	0.7670	0.2830	0.8670
14	Es9	a	1.0000	0.8330	0.5670	0.7650	0.2830	0.9150
		c	0.0000	0.1670	0.4330	0.23350	0.7170	0.0850
15	Es10	a	1.0000	0.9170	0.5670	0.7830	0.2850	0.9170
		b	0.0000	0.0830	0.4330	0.2170	0.7150	0.0830
16	Es14	a	1.0000	1.0000	0.7330	0.0000	0.5830	1.0000
		b	0.0000	0.0000	0.2670	1.0000	0.4170	0.0000
17	Fh1	a	1.0000	0.0000	0.0000	0.0250	0.0000	1.0000
		b	0.0000	1.0000	1.0000	0.9750	1.0000	0.0000
18	Gc	a	1.0000	1.0000	1.0000	1.0000	1.0000	1.0000
		b	0.0000	0.0000	0.0000	0.0000	0.0000	0.0000
19	Hbb	a	1.0000	0.3170	0.7830	0.4250	0.0000	1.0000
		b	0.0000	0.6830	0.2170	0.5750	1.0000	0.0000
20	Mup1	a	0.0000	0.0000	1.0000	0.0000	0.0350	0.9420
		b	1.0000	1.0000	0.0000	1.0000	0.9650	0.0580
21	Svp1	a	1.0000	0.5670	1.0000	0.4000	0.7170	1.0000
		b	0.0000	0.4330	0.0000	0.6000	0.2830	0.0000

*Bold numbers denote stock-specific antigens.

TABLE 3 Frequency (%) of Rats Showing Positive Reaction with RT1 Alloantisera

Rt1 Antigens	Jcl:Wistar	Crj:Wistar	Iar:Wistar	Jcl:SD	Crj:SD	Donryu
Ak	0	0	0.42	0	0	1.00
Au	0.72	0.18	0	0.08	0.72	0
Ed	0	**0.65***	0	0	0	0
Aa	0	0.65	0	0.88	0.45	0
Ea	0	0	0	0	0	0
Al	0.75	0.72	0	0.60	0.32	0
?	0	0	0.58	0.02	0	0

*Bold numbers denotes stock-specific antigen.

TABLE 4 Average Heterozygosity of Outbred Stocks of the Rat

Outbred Stocks	No. of tested loci	No. of polymorphic loci	H: Average heterozygosity
Jcl:Wistar	21	2 (9.5%)	0.0350
Donryu	21	7 (33.3%)	0.0070
Crj:Wistar	21	9 (42.9%)	0.1130
Iar:Wistar	21	9 (42.9%)	0.2030
Jcl:SD	21	13 (61.9%)	0.2020
Crj:SD	21	13 (61.9%)	0.2510

TABLE 5 Genetic Distance Among Six Outbred Stocks of the Rat

	Jcl:Wistar	Crj:Wistar	Iar:Wistar	Jcl:SD	Crj:SD	Donryu
Jcl:Wistar	—	0.237	0.326	0.402	0.414	0.277
Crj:Wistar		—	0.261	0.227	0.163	0.336
Iar:Wistar			—	0.324	0.232	0.160
Jcl:SD				—	0.151	0.474
Crj:SD					—	0.446
Donryu						—

that they have thought for a long time that they could control genetics of the colony through careful breeding schemes. However, as shown in this study, caesarian section produced a bottleneck effect on Jcl:Wistar, and Iar:Wistar is genetically different from Crj:Wistar. Thus, subcolonies exist. The only method for discriminating subcolonies from each other is by genetic testing.

For closed colonies, we propose a monitoring method as follows:

A. Testing methods

The method of the ICLAS Monitoring Center is recommended. We should use the same items for monitoring to facilitate genetic evaluation of each colony.

B. Monitoring procedures

- Genetic profiling may be performed once at the beginning of periodic monitoring and should be repeated every several years.
- Monitoring profiling should be carried out periodically (for example, once a year) using a set of markers selected among the genetic profiling markers shown in Table 2.
- Testing numbers should be done, randomly selecting from a production colony with a requirement of 50 to 60 animals per colony. If a breeder has several facilities producing the same stock, all stocks should be tested.

REFERENCES

Lindsey, J. R. 1979. Historial foundations. In H. J. Baker, J. R. Lindsey, and S. H. Weisbroth, eds. The Laboratory Rat. Vol. 1. Biology and Diseases. Academic Press, New York.

Morse, H. C. III. 1981. The laboratory mouse—A historical perspective. In H. L. Fosters, J. D. Small, and J. G. Fox, eds. The Mouse in Biomedical Research. Vol. 1. History, Genetics, and Wild Mice. Academic Press, New York.

Nei, N. 1987. Molecular Evolutionary Genetics. Columbia University Press, New York.

Phenotype Assessment Requires More Than a Casual Observation

Philip A. Wood
Professor, Department of Comparative Medicine
University of Alabama
Birmingham, Alabama

Genetic differences among animals can often lead to differences in phenotype. Now that we are in the era of creating animals with planned genetic differences, particularly "gene knockout" mice, we often anticipate what their phenotypes will be. Frequently, however, there may be no abnormalities or there may be unexpected abnormalities resulting from the intended genetic change. Frequently there are unexpected interactions with downstream pathways whereby the genetic background can markedly influence the phenotype resulting from a specific genetic change. There also have been gene knockout mice declared as having no abnormal phenotype; but when subsequent more specialized analyses were completed, striking abnormal phenotypes were discovered. Not only will genetic background significantly affect the phenotype of any given gene mutation as discussed by others at this meeting, but common environmental influences such as diet or cryptic infectious disease may also have a profound influence on the overall phenotype. The goal of this paper is to discuss a general approach for carefully assessing the many important influences on phenotype that are not often readily apparent at first glance.

It seems to me that the issue of phenotyping genetically altered animals is so complex, and subject to so many subtle factors within the animal as well as its environment, that we must begin thinking in terms of paradigms. I describe here a paradigm to consider when approaching phenotype assessment of mice and rats. This paradigm is offered as an approach undergoing further refining as our assessment tools improve. I have divided this systematic approach into primary and secondary levels of assessment for the simple reason that all possible analyses are not practical for any animals. Additionally, the primary level assessment can and should be available for investigators at most biomedical research institutions, but the secondary level of assessment will likely require more specialized expertise and equipment and could be integrated into nationally based networks established for phenotype assessment. Both components will be crucial for fully assessing phenotypes and fully using the vast number of rodent models currently being developed and studied.

PRIMARY LEVEL ASSESSMENT: FIND ABNORMALITIES

The goal of the primary assessment is simply to find abnormalities, through the following:

1. Clinical Assessment: Many knockout mice are initially on C57BL/6 × 129/Sv hybrid background, therefore controls should be littermate controls with a similar mixed background.
   a. Litter size: number born/weaned, sex, and genotype distribution
   b. Visual observation, particularly during the dark cycle when rodents are most active. Observe for behaviors that are aggressive, hyperactive, hypoactive, and so forth
   c. Observe for any coat color differences, skeletal or other body conformational changes, and failure to thrive.
2. Pathologic examination: Recommend evaluating both weanlings and retired breeders
   a. General necropsy to observe for any gross lesions and histopathology of all organs by an experienced rodent pathologist
   b. Microbiologic/serologic/parasite evaluation to detect any background infectious disease that may confuse the phenotype resulting from a gene mutation.
   c. Clinical pathology measures such as blood counts and simple urine analysis for protein and glucose.
   d. Determine life span and reevaluate phenotypes in old age.

SECONDARY LEVEL ASSESSMENT: EVALUATE AND QUANTIFY ABNORMALITIES

The goals of the secondary assessment are to evaluate and quantify the abnormalities found during the primary assessment. This will often require more specialized expertise and technology.

1. *Embryologic evaluation*
   a. If abnormal litter size and genotype distribution are observed, these animals should be evaluated for gestational loss versus neonatal loss.

This evaluation often requires timed matings with careful embryologic evaluation to detect the specific gestational stages when the animals die.

2. *Specialized pathologic evaluation*

   a. Specialized stains for lesions detected by standard workup

   b. Electron microscopy for cellular lesions that are not discernable at the light microscope level

   c. Further evaluation of any blood cell count abnormalities with FACS analysis of leukocytes and other more specific immunologic measures

   d. Specialized organ assessment such as specialty pathologic evaluation of heart changes, eye changes, bone changes, analyses of neuron/neurotransmitter distribution, and so forth.

3. *Specialized biochemical analyses*

   a. Metabolite analyses on blood, urine, tissue extracts for specific metabolites such as amino acids, lipids, carbohydrates in deficient or excessive concentrations. These assays may require very specialized equipment and expertise with small sample size.

   b. Enzyme or other specific protein analyses. This analysis would include not only assays that demonstrate the presence or absence of a protein, but also functional assays that may be crucial for corroborating any abnormal metabolite assays or blood cell abnormalities.

   c. Hormone analyses. This analysis can be particularly important in diabetic animals as well as those with failure to thrive, small body size, infertility, skeletal abnormalities, behavior abnormalities, or skin disease.

4. *Physiologic assessment*

   a. Pathologic evaluation may indicate organ dysfunction such as hyperplastic or hypertrophic enlargement, atrophy, or absence. Technologies are being developed to more thoroughly assess physiologic function such as miniaturized equipment that can transmit data via telemetry for these valuable physiologic measures in the awake unrestrained animal. Miniaturized instrumentation for procedures such as ultrasonography, magnetic imaging, DEXA analyses, indirect calorimetry studies, and other such devices are becoming increasingly available for these specialized measures in rodents, including those that are especially difficult in mice.

5. *Behavioral assessment*

   a. This is an important, developing area of biomedical research that will take advantage of the numerous genetic modeling approaches provided by rodents. There already are many behavioral differences observed among the inbred strains of rodents. With the gene mapping tools now available many genotype/phenotype correlations can be pursued including studies pursuing the genetic components of drug abuse and mental illness. There are knockout mice that also have abnormal behavior that need evaluation. This will require not only the current behavior assessment paradigms, but also specialized physiologic assessment with location specific brain implants and EEG type measures.

6. *Pathologic effects*

   a. Rodents have a wide range of susceptibilities to common laboratory pathogens, many of which alter biologic responses (NRC 1991). This susceptibility has been well documented for several infectious agents, pointing out at least two issues to consider: (1) To fully evaluate the phenotype of a new model, it should be documented free of these pathogens that may induce unwanted phenotypes. (2) For genetic manipulations involving immune functions and related effects, the animal's susceptibility to even opportunistic pathogens may markedly influence the phenotype. Thus, when evaluating the phenotype with a specific intended effect on these systems, this potential influence must be carefully controlled for and assessed.

ENVIRONMENTAL INFLUENCES

1. Dietary changes and unsuspected constituents or deficiencies in diet may play important roles in expression of phenotypes. Although many rodent diets appear fairly similar, subtle changes in constituents can have significant effects on the animals. Some examples include the possible roles that phytoestrogens, found in soybean-based protein, may have in masking gender-specific phenotypes. Likewise, studies involving blood pressure evaluation may be significantly affected by simply changing rodent diet vendors who supply different quantities of salt in their diets.

2. Significant alterations in reproductive phenotype can result with major changes in animal room temperature, humidity, pheromone effects, and noise.

EXAMPLES

1. BALB/cByJ versus BALB/cJ mice. This example will illustrate the drastic behavioral differences and metabolic differences in mice that appear very similar at first glance (Wood and others 1989).

2. Male-specific heart changes seen in mice with long-chain acyl-CoA dehydrogenase deficiency. A concern is that the soy-based protein in the diet consumed by these mice may mask the net cardiac changes seen in this model (K. B. Cox, D. M. Kurtz, and P. A. Wood, unpublished results).

3. Strain specific responses to *Mycoplasma pulmonis* infections in mice (Cartner and others 1996).

4. Examples of the unsuspected changes in arginine metabolism in rat models used in studies of salt-sensitive hypertension (Wood and others 1998).

In summary, phenotypic assessment is an important part of the genotype/phenotype correlations that we are all interested in understanding with mutant rodent models in biomedical research. It is important to follow a systematic approach for the assessment, so that the model can be used to the fullest extent. Considering the problem of trying to evaluate the myriad of effects resulting from a single genetic change is a daunting task. The goal of this presentation is to provide a framework to consider this important problem.

REFERENCES

Cartner, S. C., J. W. Simecka, D. E. Briles, G. H. Cassell, and J. R. Lindsey. 1996. Resistance to mycoplasmal lung disease in mice is a complex genetic trait. Infect. Immun. 64:5326-5331.

NRC [National Research Council]. 1991. Infectious Diseases of Mice and Rats. A report of the Institute of Laboratory Animal Resources Committee on Infectious Diseases of Mice and Rats. National Academy Press, Washington, D. C. 397 pp.

Wood, P. A., B. A. Amendt, W. J. Rhead, D. S. Millington, F. Inoue, and D. Armstrong. 1989. Short-chain acyl-coenzyme A dehydrogenase deficiency in mice. Pediatr. Res. 25:38-43.

Wood, P. A., D. A. Hamm, P. Y. Chen, and P. W. Sanders. 1998. Studies of arginine metabolism and salt-sensitivity in the Dahl/Rapp rat models of hypertension. Mol. Genet. Metab. 64:80-83.

Genetic and Phenotypic Definition of Laboratory Mice and Rats / What Constitutes an Acceptable Genetic-Phenotypic Definition

Muriel T. Davisson
Senior Staff Scientist, The Jackson Laboratory
Bar Harbor, Maine

GENETICALLY DEFINED MICE

Although "genetically defined" is often equated with inbred strains, a genetically defined strain is any strain in which the genetic background is known, is similar or identical from one mouse to another, and can be faithfully reproduced over time. Genetically defined mice are important for basic and biomedical research. They provide reproducible systems that enable investigators to replicate experiments and enable different scientists to use genetically similar or identical research animal models. This presentation discusses key elements in the use of genetically defined mice: genetic standardization, standard genetic nomenclature, genetic definitions of different types of strains, and the value and uses of different types of genetically defined strains. The focus of this presentation is the laboratory mouse, *Mus*. Mouse models are surrogates for human conditions, but they need not precisely replicate a human disease to be of value for biomedical research. More importantly, a model should be genetically defined so that the results observed can be attributed to the gene or genes being studied and the experiments can be replicated.

GENETIC STANDARDIZATION

Genetic standardization means simply that a related group of individuals can be genetically described, are similar to each other, and can be recreated by a standard and defined breeding protocol. The value of genetically standardized models is that they can be repeatedly reproduced simply by breeding. A model system is of little value unless it can be propagated reliably.

Such a model assures the continued availability of the same model to different investigators at different institutions over long periods of time. Experiments can be replicated for verification of data and experiments in one laboratory or repeated in another with the expectation that results will be similar if they are due to the mutant gene or genes being studied. It is critical to state in publications the genetic background of the mice studied so other scientists can repeat your experiment.

One of the reasons for concern about genetic definition is that with the recent strong emphasis on training in molecular biology, scientists often have an inadequate understanding of whole animal biology, classical genetics, breeding mice, maintaining strains, and keeping pedigree records or a lack of appreciation for why it makes a difference. Much of the literature published on targeted mutation mice and transgenic mice is compromised by lack of a clear definition of the genetic background on which the mutation was studied.

It is well known among mouse and human geneticists that genetic heterogeneity can alter the phenotypic expression of identical mutant genes in different individuals. This phenomenon is thought to contribute to much of the variability among human beings with the same genetic disease. Thus, it is important to keep the genetic background as homogeneous as possible when trying to determine the effects of a mutated gene in a model system. Inbred laboratory mice provide the possibility to do this. Individual mice within an inbred mutant strain are essentially genetically identical to each other except for the mutant gene being studied. Differences between mutant and nonmutant (control) mice can be attributed to the mutant gene with a high degree of certainty.

Different mouse strains are known to have different behavioral and phenotypic characteristics. Different strain backgrounds can alter the phenotypic effects of individual major genes. There are many examples of spontaneous or targeted mutations producing different phenotypes when they are transferred from one genetic background to another. If the strain used is not genetically defined, one cannot really know what aspects of the phenotype being studied are due to genetic background effects or to the mutation itself.

GENETIC NOMENCLATURE

Standard genetic nomenclature provides unique identification for different strains. Investigators reading a paper can obtain the appropriate animals to replicate the experiments described or carry out related experiments in the same system. The strain symbol also conveys basic information about the type of strain or stock used and the genetic content of that strain. Examples of symbols for different types of strains are given in the next section, describing the values of different types of genetically defined strains. Rules for symbolizing strains and stocks have been promulgated by the International Committee on Standardized Genetic Nomenclature for Mice since the early 1950s. The rules are available on-line from the Mouse Genome Database (MGD; http://www.informatics.jax.org) and were most recently published in print copy (Lyon and others 1996). Strain symbols typically include a Laboratory Registration Code (Lab Code). The first Lab Code appended to a strain symbol identifies and credits the creator of the strain. The Lab Code at the end of a strain symbol indicates the current source for obtaining mice of that strain. Different Lab Codes appended to the same strain symbol distinguish sublines and alert the user that there may be genetic divergence between the different sublines. For example, CBA/J is known to have genetic differences from CBA/CaJ. Lab Codes are assigned from a central registry to assure that each is unique. The registry is maintained at the Institute for Laboratory Animal Research (ILAR) at the National Academy of Sciences, Washington, D.C. Lab Codes may be obtained electronically at ILAR's web site (www4.nas.edu/cls/ilarhome.nsf).

DEFINITION AND VALUE OF DIFFERENT KINDS OF STRAINS

Inbred strains are defined as having been created by more than 20 generations of sibling or filial matings (symbolized F20). In reality, a strain will not be completely homozygous at all loci until it has been propagated for more than 40 generations because residual heterozygosity still can be detected at 40 generations (Fox and Witham 1997). Individuals of an inbred strain are considered to be genetically identical, and phenotypic variations are due to environmental differences. Because individuals are genetically identical, studies can be done with relatively small sample sizes. Inbred strains are valuable to define the genetics of traits such as susceptibility to infectious disease or response to specific drugs. Inbred strains that differ in such traits can be crossed together to define the genetic basis of the differences and to determine the number and chromosomal location of genes involved. They also are used when multiple genetically identical animals are needed to test the effects of a treatment. Inbred strains are typically symbolized by a few capitalized Arabic letters followed by a forward slash and a subline number and/or Lab Code. For example, CBA/CaJ is the subline of CBA inbred strain maintained first by Carter (Ca) and now by The Jackson Laboratory (J).

One should be aware that conclusions drawn from studying a single inbred strain apply only to that inbred strain. For example, if you study response to treatment with some agent, your results are specific only to that particular inbred strain. You cannot generalize across many inbred strains. If you want to model a noninbred human population, you might want to use an outbred mouse population of some kind and look for variability in response to your treatment. Because some variation will be due to genetic variability, larger sample sizes are required than for experiments with inbred mice. A caveat about outbred mice, however, is that mice coming from small closed colonies are often not as outbred as we think they are. Also, a stock that has been rederived to improve its health status has been through a breeding bottleneck that will reduce the heterogeneity in the stock.

Hybrid mice are made by crossing mice of two inbred strains together. The resulting F1 hybrids are genetically identical because at each gene they all carry both alleles from the two inbred parents. Their uses are similar to inbred strains but they are more robust. F1 hybrids cannot be self-propagated and must be created each time by mating mice from two inbred strains. Hybrid mice are symbolized using abbreviations for the parental strains. Their symbols, when correctly written, indicate the sex of each parent. For example, a B6D2 F1 hybrid is created by mating a C57BL/6 (B6) female to a DBA/2 (D2) male; a D2B6 F1 is created by mating a DBA/2J female to a C57BL/6J male.

Inbred mutant strains are inbred strains that carry one or more spontaneous or induced single gene mutations. Such strains differ from the parental strain only by the mutated gene (and in some cases closely linked genes; see below). They are valuable for understanding the effects of single gene mutations and for cloning disease genes. Differences in phenotype between mutant mice and control littermates or same strain control mice can be attributed to the mutated gene. There are two kinds of inbred mutant strains. Coisogenic mutant strains are the original strains on which the mutations occurred, and mutant mice differ from control mice only by the mutant gene. Congenic mutant strains carry a mutation that has been backcrossed onto the strain background from another strain or noninbred stock background. The nonrecombinant DNA around the mutation is from the original donor strain. This distinction is important when positionally cloning genes because in the congenic strain, differences in any genes considered candidates for the mutation may be polymorphic differences transferred with the mutation from the original strain.

Mixed inbred strains are inbred strains that are recently derived from two inbred genomes. A common example would be when targeted mutation strains are derived by sibling matings starting with the chimeric founder, composed of cells from the 129 embryonic stem cells and the host, typically C57BL/6J, mated to a littermate. Such a strain is designated using the abbreviations for the two parental "strains" separated by a comma, such as B6, 129.

In a **segregating inbred** strain, the mutation is maintained with forced heterozygosity by intercrossing heterozygotes or mating heterozygotes × homozygotes. In either case, both mutant and control animals are present within the same strain. In homozygous mutant strains, wild-type mice of the same background strain must be used as controls.

Recombinant inbred (RI) strains are sets of inbred strains created from sibmated F2 progeny produced by crossing mice from different inbred progenitor strains, such as C57BL/6J and DBA/2J. RI strains are valuable for mapping phenotypic or quantitative traits that differ between the progenitor strains. They are especially valuable for controlling for environmental variability in a trait because several genetically identical mice from each line in a set can be typed to score the line for a trait. Crossover events can be detected by strain distribution patterns (SDPs) of alleles among the RI lines, typically using a series of regional markers (Bailey 1971; Taylor 1989; Mouse Genome Database) (Table 1). RI strain sets are like a linkage cross-frozen in time, and genotyping is cumulative.

Recombinant congenic strains are sets of inbred strains derived in a similar manner to RI sets except that one or more backcrosses to one parental (designated the background) strain are made after the F1 generation, before inbreeding is begun. The other parental strain is designated the donor. The proportion of background and donor genomes is determined by the number of backcrosses preceding inbreeding. Because these sets are typically constructed after two backcrosses, each recombinant congenic strain usually contains approximately 87.5 % of its genes from the background strain and approximately 12.5% of its genes from the donor strain (Moen and others 1991; Stassen and others 1996). As with recombinant inbred strains, a detailed characterization of SDPs of genes within a strain set may be used to determine linkage relationships between loci and chromosomal segments associated with a trait such as tumor susceptibility. Typing of recombinant congenic strains is useful in the analysis of complex genetic traits in the mouse (Moen and others 1991).

Congenic strains are derived by successive backcrosses in which one strain (the donor) donates a segment of chromosome to the recipient (background or host) strain. Congenic strains are genetically almost identical to the background strain except for a short chromosomal segment contributed by the donor strain. The most familiar congenic strains are histocompatibility congenics (Snell and Bunker 1965).

More detailed information on strains of laboratory mice may be found in The Jackson Laboratory's *Handbook on Genetically Standardized JAX Mice* (1997).

GENETIC MONITORING

Genetic monitoring is critical to maintaining genetically defined strains. Although this topic is covered in another presentation, I touch briefly on it here. The best protection against genetic contamination is good animal husbandry and record keeping. There is no substitute. Genetic monitoring is just what its name describes—monitoring to assure that mistakes have not been made. There are two kinds of genetic monitoring: (1) genetic background monitoring to detect and eliminate possible genetic contamination, and (2) mutation monitoring to assure that the mutation carried by a mutant strain is still present. Neither is particularly difficult to do, but both are crucial to ensure strain integrity and, in the case of mutant strains, avoid the loss of valuable mutations. Genetic background monitoring is typically done by screening a set of biochemical and DNA markers in progenitor breeding pairs, that is, the breeding pairs in each, or at least every other, generation that are in the straight line pedigree for the strain. It also is wise, especially in large colonies with several generations of expansion, to monitor mice chosen randomly from the expansion colony. Because this type of monitoring is retrospective, a change might not be detected until it is widespread. One needs only a minimum set of markers whose allele distribution distinguishes that strain from others in the same mouse room. Coat color is a simple visible marker that requires no genotyping. Its use can be enhanced by simply interspersing strains of different coat colors in the same mouse room. In addition, genetic contamination between different strains usually results in a sudden increase in reproductive performance. Be suspicious if mice from a strain with low reproductive performance suddenly start to breed well.

Monitoring spontaneous mutations can be as simple as visually observing mice in each generation for the mutant phenotype. Although this is usually adequate, if the phenotype is common to multiple nonallelic mutations, there is a risk that one mutation may be lost and replaced by another in a segregating mutant strain. For example, all mutations that affect the cerebellum cause very similar balance defects. Once a spontaneous mutation is cloned, it can be followed in nonaffected carriers or verified periodically by DNA genotyping. Targeted and induced mutations and transgenes also can be monitored visually if there is an associated phenotype. For those in which the mutants die during gestation, DNA genotyping may be used to follow the mutated gene or transgene. If several targeted mutations have been created using the same type of construct, the same polymerase chain reaction (PCR) protocol can be used to economically genotype mice of all strains. For example, a set of targeted mutation strains made with *neo*-containing targeting vectors can be typed simultaneously PCR-typing for *neo*. However, one should periodically genotype each strain with allele-specific markers to ensure against cross-contamination between such strains.

GENETIC DATABASES

Several databases are available that have information on genetically standardized mice. The Mouse Genome Database mentioned above has a list of inbred strains of laboratory mice, as well as information on mouse genomics and gene expression. The strain list is prepared by Dr. Michael Festing of England, who also is responsible for assigning strain names to new inbred strains. The Laboratory Registration Code database is maintained at ILAR, also mentioned above. The Jackson Laboratory web site (http://www.jax.org) has two databases that list targeted mutation and transgenic mice: (1) TBASE, the transgenic database, which was developed by Dr. Rick Woychik, was transferred in 1998 from The Johns Hopkins University; and (2) the Induced Mutant Resource (IMR) database, which lists induced mutant strains available from The Jackson Laboratory. Both are supported by the National Center for Research Resources (NCRR).

Scientists using mice also should know how to find other species' databases because it is important to try to give homologous genes in different species the same or similar symbols. Some databases for species most commonly referred to in comparative studies are listed in Table 1.

TABLE 1 Selected Genetic and Strain Databases Available on the World Wide Web (WWW)

Site	Contents	Web address (URL)
Mouse Genome Database	Mapping data (all techniques) genetic, cytogenetic, physical, and comparative mapping data	http://www.informatics.jax.org
MRC[a] Mammalian Genetics Unit	Comparative maps, strain list	http://www.mgu.har.mrc.ac.uk/
The Whole Mouse Catalog (formerly Mice and Rats Home Page)	Links to web sites for mouse and rat research	http://www.rodentia.com/wmc
Animal Genome Database in Japan	Mouse genetic mapping data, cytogenetic maps	http://ws4.niai.affrc.go.jp/
Human Genome Database[b]	Human gene symbols	http://bioinfo.sickKids.on.ca/ http://gdbwww.gdb.org/
Human Gene Nomenclature Database	Human gene symbols	http://www.gene.ucl.ac.uk/ cgi-bin/nomenclature/ searchgenes.pl
National Center for Biotechnology Information (NCBI)	Mouse/human comparative maps, links to other databases	http://www.ncbi.nlm.nih.gov/ Homology/
Rat Genome Database	Rat genetics	http://ratmap.gen.gu.se/
Roslin Institute Bioinformatics	Pig, sheep, cattle, chicken	http://www.ri.bbsrc.ac.uk/ bioinformatics/
FlyBase	*Drosophila* genomics	http://flybase.bio.indiana.edu http://shigen.lab.nig.ac.jp:7081
Zebrafish Informatics	Zebrafish genomics	http://zfish.uoregon.edu/ZFIN/

[a]MRC, Medical Research Council.
[b]Note: At the time of this writing, the Human Genome Database is in transition between the two sites listed.

TRAINING SCIENTISTS TO USE GENETICALLY DEFINED MICE

Finally, I would like to return to the point that many scientists trained in the 1980s and 1990s have not really been trained in practical genetics or animal husbandry. With the current research trend moving back toward phenotype analysis, mutagenesis, and whole animal studies, there is a desperate need to provide programs that train scientists to understand, work with, and maintain genetically defined mice. We need resources to provide training in practical genetics, breeding schemes, record keeping, and mouse husbandry. I think an increasing number of investigators today recognize the effects of genetic background on phenotype and the importance of using genetically defined strains; however, many need resources to help them with the practical aspects of creating and using such strains. For example, The Jackson Laboratory has an annual course called Experimental Genetics that is geared to graduate students, postdoctoral fellows, and investigators changing their research programs to use mice. The course teaches practical Mendelian genetics, how to breed animals, how to keep records to avoid mixing up mice within the colony, and basic animal husbandry. Unfortunately, the course handles only about 30 students a year and, as far as I know, is the only course of its type in this country. We need more of this sort of course introduced into graduate schools or offered in training programs similar to that at The Jackson Laboratory.

REFERENCES

Bailey, D. W. 1971. Recombinant-inbred strains. An aid to finding identity, linkage, and function of histocompatibility and other genes. Transplantation 11:325-327.

Fox, R. R., and B.Witham, editors. 1997. Handbook on Genetically Standardized JAX Mice. The Jackson Laboratory, Bar Harbor, Maine.

Lyon, M. F., S. Rastan, and S.D.M. Brown. 1996. Genetic Variants and Strains of the Laboratory Mouse. Oxford University Press, Oxford.

Moen, C. J., M. A.van der Valk, M. Snoek, B. F. van Zutphen, O. von Deimling, A. A. Hart, and P. Demant. 1991. The recombinant congenic strains—A novel genetic tool applied to the study of colon tumor development in the mouse. Mamm. Genome 1:217-227.

Snell, G. D., and H. P. Bunker. 1965. Histocompatibility genes of mice. V. Five new histocompatibility loci identified by congenic resistant lines on a C57BL/10 background. Transplantation 3:235-252.

Stassen, A. P., P. C. Groot, J. T. Eppig, and P. Demant. 1996. Genetic composition of the recombinant congenic strains. Mamm. Genome 7:55-58.

Taylor, B. A. 1989. Recombinant inbred strains. Pp. 773-796 in M. F. Lyon and A. G. Searle, eds. Genetic Variants and Strains of the Laboratory Mouse. 2nd edition. Oxford University Press, New York.

Genetic and Phenotypic Definition of Laboratory Mice and Rats / What Constitutes an Acceptable Genetic-Phenotypic Definition

Joseph DeGeorge
Associate Director, Pharmacology and Toxicology
ORM, CDER/FDA
Rockville, Maryland

NECESSITY FOR GLOBALLY STANDARDIZED OUTBRED RATS FOR CARCINOGENICITY BIOASSAY

• Why is the issue of outbred strains of current importance from a regulatory perspective?
• Why was there virtually no interest in the issue several (for example, 10) years ago?
• What makes outbred strains important now?

I believe the answer to all three questions can be summed up in one simple sentence: There has really been a large change in the paradigm of pharmaceutical development. And that is the basis for my concern today.

CHANGE IN LABORATORY ANIMAL SCIENCE

The change has occurred over the last 10 to 15 years and mainly over the last 5 to 10 years. I address below some specific aspects of the changes that have led to my concern about standardization of the animal models on which we rely. One additional point to keep in mind is that as a regulatory agency, the US Food and Drug Administration (FDA) is the end user of the data from all of the pharmaceutical testing that goes on. We have to rely on it to make judgments about potential human health risks. The outcome is not a research paper but rather, marketing to millions of people around the world of a product that has gone through a particular testing process.

GLOBALIZATION OF DRUG DEVELOPMENT

The first of the major changes that have occurred recently in pharmaceutical development is the globalization of the drug development process. It has been mentioned that there is an International Conference on Harmonization of Technical Requirements for International Registration for Pharmaceuticals for Human Use (ICH), where we have harmonized and in fact have agreed on certain standards. However, beyond that is the fact that pharmaceutical companies are almost no longer national. There are very few national pharmaceutical companies, and most market worldwide. Most also develop their drugs as a worldwide activity. Worldwide marketing, in fact, has reached the point where we are developing a common technical document so that the test studies—the same studies—are being submitted simultaneously around the world for marketing approval. This is an attempt to achieve drug approval in Europe, the United States, and Japan at roughly the same time.

The extent of globalization is very large, and "global" companies are headquartered throughout the world. There is a perceived need for global standards that companies can follow to make certain that if they do a study for one country, it is acceptable in another country. This standardization is actually the basis for Office of Economic Cooperation and Development (OECD) guidances, which is familiar to most of you as providing information on various test paradigms and specifically for pharmaceuticals. ICH guidance refers to specific aspects of carcinogenicity testing, genotoxicity testing, and reproductive toxicity testing for pharmaceuticals. If companies follow these guides, no matter where in the world they do the study, that study is accepted internationally—at least within the United States, Europe, Japan, Australia, Canada, Taiwan, and wherever else drugs are manufactured. Companies are looking to these standards to try to establish test systems and hopefully market a pharmaceutical.

One of the accomplishments of this harmonization of standards has been the elimination of duplicate testing. Thus, when an institution or company proposes a drug development plan for Japan, it is not necessary to complete another set of tests for the United States. In the past, that necessity might have been the case; or they would have made certain that the studies were done in both the United States and Japan. Furthermore, this harmonization has actually eliminated many specific national test requirements. The US test requirements do not differ from the Japanese or the European under this ICH process, at least for pharmaceuticals. Such standardization is one aspect of this global drug development plan that has an impact on the use of animals.

SEGMENTATION OF TOXICOLOGY TESTING

Another aspect of globalized pharmaceutical development is the segmentation of toxicology testing. International companies have, in fact, often changed their development processes. It is often no longer an all in-house operation done in a particular facility where all the data are generated on the same colony of animals, which then undergo every test. There is an increasing use of contract facilities, a blend of contract facilities with sponsor facilities, and an increase of multisourcing outsourcing, going to the lowest bidder. In other words, if it is cheaper to do a certain study in the southwestern United States and another study more cheaply in Japan, they will do those studies in those two places, presumably using the same animal models and the same strain.

One real example is a company that conducts its chronic toxicity studies in their European facility, dose-ranging studies for carcinogenicity studies in their US facility, and then contracts out the carcinogenicity studies. Looking at the results from all those studies from different sources, we have to make a decision about how those results from the chronic study and the dose-ranging study apply in the interpretation of the carcinogenicity study. If the animals appear to be responding differently, we have a big problem. That, then, is what I mean by segmentation of toxicology testing and how it is a major issue in terms of international drug development.

A major concern in terms of carcinogenicity testing—and one reason that we are focusing now on rats rather than mice—is because (as Dr. Usui mentioned, according to the ICH guidance) our long-term carcinogenicity studies are now generally performed using rats rather than in mice. We are using rats mainly because transgenic models are available primarily in mice; so to have two-species testing, the standard 2-year bioassay tends to be done in rats. Another reason is that most pharmaceutical development strategy uses rats as the rodents and dogs as the nonrodents. Because these companies develop a large database on the effects of the pharmaceutical on the rat during their testing, they want to be able to use that information in test approach selection, dose selection, and interpretation of results.

The preceding observations lead to a couple of concerns in terms of carcinogenicity assessment, the first of which are differences in stock survival. Dr. Usui mentioned that these differences might be related to diet. In some cases, they may be related to housing conditions; some companies let their rats get very fat because they put the food in wire baskets and allow them to feed ad libitum. Other companies and facilities put their food in jars and as the rats get fat, they can no longer reach the bottom of the jar. The rats thus undergo a kind of spontaneous dietary restriction, which maintains their weight at a lower level; and those differences, although not necessarily outbred related, can affect the outcomes and interpretations of studies. Diet and body weight are two factors that can clearly affect survival characteristics of the strain and the species. If they are overlaid with differences in survival within a strain of the outbred animals, then you end up with a very difficult problem for interpretation of carcinogenicity data.

Diet, weight, and survival characteristics can also result in differences in responsiveness to various pharmaceutical products. A set of similar studies with outbred animals may show a difference in response to treatment at some dose level. For example, if a chronic toxicity study is performed in a contract facility by gavage, and a second study for dose ranging is done in the in-house facility by dietary administration using dietary restriction (because it is much less expensive to plan the carcinogenicity study by dietary administration), and the studies result in different outcomes regarding the toxicity of the product, one will not know whether those differences are related to the modalities of the ingestion of the drug, the husbandry of the stock being treated, outbred differences from different sources, dietary restriction, or some other factor.

NEED FOR INTEGRATED FINDINGS

When we interpret carcinogenicity studies, we no longer can simply state, "It was positive in this site, and it was negative in all of the other sites." As a part of the ICH guidance for pharmaceuticals (with which the US Environmental Protection Agency [EPA] agrees), we now perform integrated assessments of carcinogenicity—weight of the evidence. It is necessary to consider all of the data collected on the product and try to make a determination of whether or not those findings are important for human risk. In so doing, it is often necessary to rely on historic data to interpret the findings. When looking at historic data, it is important to note its relevance to the particular strain of animal and, unfortunately, to the animal stock.

MANAGING CHANGES OVER TIME

For some facilities that have been testing for 20 years, changes over time can clearly be seen in the historic response rate for spontaneous tumor incidence. Other facilities have no historic data but instead, rely on published data. How relevant is that published data to that in-house-contained and bred outbred or inbred strain? The answer to such a question is critical for the assessment and determination of the carcinogenic risk for humans. It is imperative for us to understand and manage test data accurately.

One approach to managing test results is to try to control as many variables as possible. That method is the best solution—to try to control the dietary supplies, the strain of animals, the stock of the animals, the dosing regimens, and the like so that one can say that the corresponding data are reliable for interpretation. If a second person repeats the experiment or study, the person can be reasonably confident of obtaining the same result. However, controlling all of the variables is probably not feasible in the global development arena. How can you ensure that all animals have the same diet, no matter where the test facilities are located? I believe we can at least begin in this direction by controlling stocks of animals.

Nevertheless, we must be very careful not to select a single stock of a single strain of animal and proceed with that alone. It is known that some animal strains metabolize and respond to pharmaceuticals more like humans than do other strains. Having those strains for testing to find the most relevant response is important so that resulting data from a well-controlled strain is relevant to humans. If we pick only one stock and standardize it, we will loose a great deal because even though we will have a reproducible result with that strain, we may not want to use that strain for much of our testing.

CIEA/NCRR/NIH Genetic and Microbiological Monitoring of Mouse and Rat Resources: Directions for the Future

Tatsuji Nomura
Director, Central Institute for Experimental Animals
Kawasaki, Japan

During the US/Japan Meeting about 15 years ago, we decided to select specific microbiological items based on criteria that determine minimum requirements. As Dr. Itoh explained, these requirements correspond to those of the International Council for Laboratory Animal Science (ICLAS).

The ICLAS concept of microbiological monitoring is the same as that for genetic monitoring, that is, monitoring of animals that have been genetically controlled (such as inbred, hybrid, or congenic animals). Microbiological quality monitoring is applied only to microbiologically controlled animals that are barrier sustained, such as gnotobiotes and SPF animals.

DIFFERENCES BETWEEN COUNTRIES

In the United States and Europe, there is no microbiological monitoring; however, there is health monitoring. I can understand this concept because, from a practical standpoint, disease is the most important in those countries. However, health monitoring does not cover the microbiological quality of the animals. In addition, we have had recent experiences with new users—molecular geneticists—who have asked for more sophisticated high-quality mice for analysis of the expression of the introduced gene. For these users, we believe that it is important to monitor the microbiological quality of the animals.

It is our desire to reach a consensus regarding worldwide minimum requirements for genetic and microbiological monitoring. Each country has its own requirements for microbiological monitoring, which can be added as options; but we should at least establish minimum requirements.

QUESTIONS AND ANSWERS

T. GILL: I believe the critical point in monitoring is not with people who understand (such as with all of us, who agree), but with the political process. Unless a lot of people make a lot of noise, nothing is going to happen. If you publish a set of standards, the microbiological and genetic standards must be checked. If you promulgate the standards among the scientific community, then when scientists ask for animals they will ask the suppliers, "Do you fulfill these criteria? If you don't, we won't buy animals from you." Therefore, I think that a relatively simple list of microbiological requirements and a simple list of genetic requirements that correspond with each animal should be developed, and the scientific community should be encouraged not to buy from suppliers who do not provide this information.

T. NOMURA: What you describe is happening in Japan. The pharmaceutical industry will never buy from a breeder who has no test results from the ICLAS Monitoring Center. This has made the quality of breeders very high. As I mentioned, the United States should have a reliable, neutral, authorized monitoring center for microbiology and genetics on which everyone can depend.

T. GILL: The key point is that an institution independent of the animal suppliers has the authority to apply pressure from the pharmaceutical users to maintain these standards. That is why assigning the implementation of standards to the suppliers will never work because increasing profit is the basic drive of the suppliers. Sometimes high quality and increasing profits do not exactly mesh. Consequently, an outside group has to apply this pressure, and I think the users are the outside group that has to do this.

T. NOMURA: ICLAS is the only neutral organization that can do that.

T. ALLEN: You bring up a problem that is a real issue at NIH. Even though there were rules everywhere that you could not do what I am going to describe, it still happened. Within 6 months after we were finished with the ectromelia outbreak in 1979 to 1980, one of my colleagues on an airplane sat next to a man who had in his pocket a mouse that he was bringing in from the same institution where the outbreak originated to get around the rules and regulations. Asking staff to spend $1000 for a map test when they are on a budget is really a major problem.

T. NOMURA: I would like to mention again that the microbiological monitoring requirements that we selected are based on the contamination map of Japan. We have checked and know what agents are currently spreading. In addition, the standards for selection have to be decided by the users, not by the breeders. In Japan, we try to focus on the most critical users—industry. This focus is especially important when they are conducting long-term 2-year toxicology studies and they ask us to check certain items, especially pathogens causing inapparent infections. We base our selection in this way.

CIEA/NCRR/NIH Genetic and Microbiological Monitoring of Mouse and Rat Resources: Directions for the Future

Neal West
Program Director, Comparative Medicine, NCRR, NIH
Bethesda, Maryland

NIH STRUCTURE

As most of you know, there really is no centralized National Institutes of Health (NIH) planning body that decides future directions for the genetic and microbiological monitoring of rodent resources, except for a very few activities such as our sponsorship of this group through the National Center for Research Resources (NCRR). The NIH structure comprises 18 institutes and six centers. Most of our institutes and centers also have extramural and intramural components, which have very different missions and functions. This structure contributes to a very diffuse governmental authority. In our society, leadership derives not just from NIH, but also from academic, industrial, and research institutes such as The Jackson Laboratory, which, of course, we support financially, providing direction and encouragement.

Our grants do have strings attached. We produce guidelines and generate initiatives. In addition, I might mention some concrete initiatives or "case studies," particularly for the benefit of our Japanese colleagues, who may not be as familiar as others with how NIH proceeds in setting policy.

NCRR'S MISSION

Priority Setting for Mouse Genomics and Genetics Resources (Dove and Cox 1998) (which we call "The Mouse Report") arose from a workshop held in March 1998 and organized by NIH Director Dr. Harold Varmus to respond to the community's needs. This workshop, also referred to as the "Dove and Cox

Workshop," was named after two of the leaders in mouse genomics and genetics. A report was generated by the members of that workshop and was put on-line on the NIH Director's home page in June 1998. The report, which describes the activities in dollars (millions of dollars), was circulated to the very broad scientific community, and expectations were very high. Everyone assumed that the money was already available; however, we still do not have the money. The process takes time, and money must be allocated in future years.

Dr. Varmus, who had initiated the original mouse workshop, set up a trans-NIH working group to which I was appointed by Dr. Vaitukaitis to be the NCRR representative. Dr. Varmus came to the group at our first meeting on June 25 and said he wanted to see the infrastructure for mouse research in the United States built up, and he gave us—people from all of these twenty-some institutes and centers—the charge: Do some inventory, prepare a spreadsheet, and explain both what we are doing and what we are going to do.

Obviously NCRR has a legitimate, historical, well-established record and mandate in the area of rodent resources, and especially mouse resources. We certainly do a fair amount of training, especially for veterinarians; and automatically this working group looked to NCRR, especially in the area of resources and training.

As I mentioned, we are already heavily committed at The Jackson Laboratory, and we have learned much from that relationship, which has formed some of the basis for our initiatives in the planning stage. On October 5, 1988, what was called a "recalibration meeting of the mouse working group" was held, and Dr. Varmus met with a number of people who were involved in the workshop and asked us to describe where we want to go and where we would apply the resources. The only real product of that meeting was the general consensus that physical mapping must come first.

Physical mapping is somewhat distant from the kind of resources and issues that are being discussed at this meeting. Nevertheless, we may have the advantage of time because many things that NCRR does (such as infrastructure and training) must come first to build an infrastructure toward future progress. Unfortunately, other emphases may siphon off or divert some of the potential resources, but Dr. Vaitukaitis has advised us to proceed. We hope to garner some additional support from other institutes and centers. In any case, we are committed to fulfill our role to serve, as Dr. Vaitukaitis has described NCRR's mission, as a catalyst for discovery.

We will try to increase training, provide the resources, and plan for the phenotyping that everyone has come to realize is going to be a large part of the activities. Although I do not wish to criticize the brilliant microbiologists, biochemists, geneticists, and structural biologists who are participating, there appears to be a sudden reality check occurring in the complexities of what is called functional genomics (which I used to call phenotyping)—how many people it takes, how much effort is required, how long it is going to take, and how to marshall those resources to make use of the ability that now exists to generate incredibly large numbers of mutant animals. These resources will not be very useful unless an infrastructure is built and includes the pathology and phenotyping needed to make sense of changes generated in very large numbers.

DATABASE RECOMMENDATION

The other part of the reality check relates to databases, which appear to be a real problem financially, conceptually, and organizationally and which sometimes appear to be an almost unsolvable issue. Everyone agrees that databases must be integrated and linked. There should be a standardized nomenclature. Needless to say, the development of databases is going to be extremely expensive. There is currently no plan to design and organize those databases. There may be others here who have a different perspective on this subject.

QUESTIONS AND ANSWERS

J. L. VAITUKAITIS: I would not feel negative about the direction and scope of the mouse working group activities. Some of those individuals are "gene jockeys," and it is only natural that they would suggest what they did.

N. WEST: I agree, although I felt that they dominated the meeting.

J. L. VAITUKAITIS: I would not worry about that at all.

N. WEST: The other very timely initiative (in which Drs. Gill, Pakes, and Nomura participated) is the initiative of the August 1998 Rat Model Repository Workshop, which recommended a national rat genetic resource center to select, maintain, distribute, and preserve genetically defined rats (at least 50 new strains per year). With regard to future directions, I am not aware of any plans for implementation of the recommended rat resource center, although there might be some partnership with industry. Details of any initiative remain to be seen.

Also included in the workshop report is the statement that the intramural NIH genetic resources are clearly inadequate to serve the growing needs of the extramural community and, in fact, NIH is not well structured to do that. Obviously, whatever emerges, given our mandate and our guidelines, NCRR will inevitably lead the rat resource initiatives (although we do not yet know in exactly what form).

Both the mouse and the rat reports address the issues discussed here today—genetic and microbiological monitoring. However, there is a fear that something may be lost in the translation of good recommendations into actual initiatives. Many things discussed here today are very important but are also very expensive. Perhaps not fully appreciated is the great contrast—the difference in sophistication—that exists among some of the policy makers and the working groups in the meetings that I have attended at NIH and all the wonderful things I hear when I sit in a group like this.

LEARNING FROM EACH OTHER

Obviously this group is incredibly sophisticated concerning the issues under discussion, and I have a fear that some details could be lost. You focused on many important considerations for the rodents and for other animals, including genetic monitoring, harmonization, husbandry, and cryopreservation. One additional issue relates to genetic drift. Cryopreservation is a very good way to maintain a stock against genetic drift (or loss, for that matter). My friends at The Jackson Laboratory have educated me well regarding how the economics of preservation may favor increased cryopreservation.

I believe that this group—certainly individually, and collectively through many areas of expertise—has much to teach the NIH and the extramural community. There are many venues, and of course there are diverse audiences. I also believe that we at NIH have much to learn from the Japanese model for supporting these initiatives through public and private sources, including support from pharmaceutical companies. The Japanese model is structured slightly differently, and NIH may be less sophisticated with its structure.

In addition, quality control standards can be raised and improved in the United States only when the public and private sectors work together closely, and I am not certain that they always have done that. I believe these US/Japan meetings will become increasingly important as the worldwide research community becomes more interactive—more scientifically integrated—and as there is more trafficking (such as in animals and embryos) and information exchange.

POLICY SETTING

To return to the local from the worldwide perspective, I think most of you know that Dr. Whitehair requested a decrease of his responsibilities sometime ago. Although he is not retiring at this time, he has requested that a new Director of the Comparative Medicine area be selected.

Dr. Vaitukaitis has recently elevated the status of the Comparative Medicine area Director's position to the policy-setting level at NIH, that is, the Senior Executive Service (SES). These SES positions are very difficult to add to an institute or center at NIH. She announced this week that John D. Strandberg, D.V.M., Ph.D., currently the head of Comparative Medicine at The Johns Hopkins University, has been appointed to that position and will start at the NCRR on January 3, 1999. Unfortunately, Dr. Strandberg is unable to be present today due to a personal emergency. Nevertheless, I believe that these rodent initiatives and this key appointment are indications that the issues of concern to this group have great visibility in NCRR and at NIH. These indications bode well for the national leadership that NIH can provide in these areas.

REFERENCE

Dove, W., and D. Cox (Summarizers). 1998. Priority Setting for Mouse Genomics and Genetics Resources. (http://www.nih.gov/welcome/directors/reports/mgenome.htm).

Closing Comments / Summary of Presentations

Thomas J. Gill III
Menten Professor of Experimental Pathology and
Professor of Human Genetics
University of Pittsburgh School of Medicine
Pittsburgh, Pennsylvania

INTRODUCTION

In the Introduction to his talk, Dr. DeGeorge presented his disclaimer as a federal official in the regulatory business. That reminds me of the situation in the Middle Ages: By volunteering to go on a crusade, you were forgiven all of your past sins and debts. I would like to take the same position now and point out that undoubtedly my summation reflects views through which I cannot help but filter what I hear. Many of these views are probably not new.

Dr. Nomura pointed out that the US/Japan meetings began in 1980 and that the Rat Genetic Nomenclature Committee was established in 1994. That committee meets in conjunction with the Rat Workshop and has dealt with nomenclature problems quite consistently. Even before 1980, however, there was a very strong relationship between science in Japan and in the United States. There was a formal US/Japan scientific exchange that dates back to the late 1960s and early 1970s, and I participated in some of the immunological exchanges under that program. So this is a longstanding relationship, and I think we have built good bridges scientifically and technically.

NEED FOR GENETICALLY DEFINED ANIMALS

Of the needs I have heard expressed at this meeting, I have noted the important need for genetically defined animals. The problem involved in fulfilling this need is not just with the generation of the animals but also with defining the animals. One aspect of this problem has been illustrated by some of the discussions that came up today, that is, the use of the words "outbred," "closed colony," and "random." I think everyone who used those words today used them differently. The standard genetic definition of an outbred colony is one that is structured genetically, specifically to maximize diversity; random bred refers to any animals that are running around; and closed colony, to animals running around in a relatively confined space. Each group is genetically very different. Unfortunately, not only is the terminology different, but also the animals are used differently. Especially in large-scale technological settings, this difference has been a major drawback.

The background of a specific genetic defect is very important. The expression of a gene is a function not only of the gene itself, but also of its background. What we see in human genetics as variable penetrance is really (at least in my opinion) the expression of a major gene modified by several modifier genes, rather than one gene simply not showing up in one setting quite as strongly as it does in other settings. This expression is another example of the importance of genetic background and the definition of genetic background in trying to perform disease-related studies.

It has been pointed out that there has been a very strong thrust in the past decade or so, when molecular biology has come to the fore, to ignore live animal studies. A number of my colleagues in the medical and scientific world have said that we basically no longer need experimental animals because with genetic tools, we can study humans and solve all of the problems—not only of disease but also of basic biology. Unfortunately, this was the funding position of the NIH for a long time, and I think one unfortunate aspect of this attitude in the NIH is that many animal resources have been let go and many developments in basic animal models have been put on the back burner. I certainly hope that the Mouse Genome Initiative and the Rat Genome Initiative will do something to change this.

DEVELOPMENT OF GENETICALLY ENGINEERED ANIMALS

Another big problem that we face today is the result of the development of genetically engineered animals. As an example, if you insert a transgene (such as HLA-B27, to study arthritis) and you find significant phenotypic variation, you must note that this phenotypic variation may be a function of where the transgene is inserted. There is generally no way to insert a transgene in a specific place consistently. Therefore, the transgenic animal having HLA-B27 in one laboratory is used the same way as the HLA-B27 transgenic animal in another laboratory, but they may be significantly different. I think also, as one continues to look at gene expression, it is necessary to look not only at the gene that has been transferred but also at the control regions that affect the expression of that gene.

If you look at studies on the beta globin genes in the human globin system, you will see that the long-range locus control region for the beta globin system is located far away from the beta globin genes themselves. So if you want to look at a genetic expression problem, you have to look not only at the genes themselves, but also at the locus control regions, which may be quite far away. This is the kind of genetics problem that must be addressed by any group trying to standardize animals for genetic research. The terms now coming into general use reflect this need: "physiological genetics" and "physiological genomics." Physiological genetics is an old term, which I believe you will find in some of the early writings of William Castle. The term has been revivified now that genomics has become a promising field. One of the major impetuses for the development of experimental animals and for putting the effort into experimental animals that is represented by our meeting today is the tremendous impact of genomics and the possibility that it will solve many physiological problems.

At the risk of alienating a number of my colleagues, I would like to point out that molecular biology is a tool. Molecular biology is no different from serology, coat color, or eye color; it is a tool to look at variation and the transmission of traits from parent to offspring—that is what genetics is. Ultimately, you cannot do these kinds of studies unless you have live animals, and I think that this point has been missed by a large part of the scientific community and, unfortunately, by a large part of the funding agencies that supports the scientific community. It is fashionable to support molecules but not animals. Now that we are swinging back toward animals, we find that we have lost great resources, many of which I think are critical to future progress.

We must develop models that are important for biomedical research and for the study of disease. This is not to say that the study of molecular biology to understand gene function is not important; however, as was pointed out a long time ago, disease is an experiment of nature, and it is a probe by which you can perturb a normal biological system. Thus, from the basic scientific point of view, you can look at a disease model as a perturbation of a normal physiological function. From the medical point of view, you can look at a disease model as a way to give you some insight into the pathogenesis of human disease that will be useful in diagnosis and treatment. As animal models are developed, I think one has to have the focus on the development of useful disease models, which, in the long run, will be useful both biologically and medically.

Because transgenic and recombinant animals and various disease models are generated "in everyone's closet," there has to be a mechanism for deciding which ones are going to be preserved and which ones are going to be lost. As a pathologist, in a world other than the scientific world in which I am now speaking, you make a diagnosis; you know that sometimes you are going to be wrong; you are never going to forget your mistakes; but you have to live with them and move on. I think the same kind of mentality has to be brought into the selection of disease models. You have to make the best judgment you can make at the time, make your selection and move on knowing you are going to make some mistakes but accepting the fact that you are going to have to live with them.

I will not belabor my well-known reservations about animal welfare and alternative medicine, but I think that looking at medical problems from the point of view of the worm or the computer is not the best approach. I unabashedly think humans are more important than worms or computers, but one cannot experiment on humans except in very restricted contexts. There are some who argue against the use of animal models who have seriously suggested that experimentation should be carried out on humans—either volunteers or prisoners. It is difficult to believe that people can make that kind of a suggestion; it is positively atrocious.

IMPORTANCE OF DISEASE MODELS

To illustrate the importance of disease models in a variety of settings, it is necessary to look not only at the importance of understanding the disease processes and their treatments, but also at the translation of research into clinical practice. Two examples follow.

You all know the famous story of Pasteur and his rabies vaccine, for which he grew the virus in rabbit spinal cord. Little Joseph Meissner was inoculated and saved from rabies. The vaccination then became standard medical practice. In later years, we find out that the autoimmune responses to the rabbit spinal cord caused demyelinating diseases in some cases, but by then it was standard medical practice. If you did not vaccinate, you could get sued; however, if you did vaccinate, you ran the risk of disease. It was not until many years later and much experimentation that the virus was grown in eggs and did not cause the autoimmune problems. The problem, however, was that the vaccine was not thoroughly evaluated prior to its use, and it took from the middle of the 19th century to the middle of the 20th century to rectify this problem.

My second example is a more current problem that has been generated by the treatment of recurrent spontaneous abortion in a population in which childbirth is being delayed until a later and later time. There are more problems with conception, more problems with fertility, and enormous social pressure to do something about fertility in the setting where it is impaired. Some studies based on one animal experiment and four patients in London proposed that immunizing women with their husbands' leukocytes to develop blocking antibodies prevented the rejection of the fetus. It became a major industry in obstetrics practice. I can assure you that the basic science is wrong; and the clinical studies (few of which were done thoroughly), when put together in a meta analysis, support the fact that this procedure is at very best marginally effective and probably not generally effective. So here is a very recent example of how something was taken on the basis of an unconfirmed animal study and put into clinical practice. It is now a tremendous problem, which we are trying to analyze and rectify. Thus, the role of good animal models and the thorough study of disease and its treatment before translation into the clinic are important.

STANDARDIZATION AND MONITORING

It is important to standardize animals for research with oversight by the scientific community. This means that the broad scientific community, with combined knowledge in the various aspects of disease, genetics, and microbiology, should oversee the development and dispensing of these standardized animals. In my opinion, the concept of standardizing animals and monitoring them has not been appreciated thoroughly. Dr. Nomura has pointed out this concept at a number of our meetings.

From the genetics point of view, it is necessary to determine whether you monitor one gene or several genes for a polygenic system; or do you monitor the genome of the whole animal. From the microbiological point of view, do you monitor a standard panel or do you monitor everything you can put your hands on; or is there a difference? How do you characterize initially and how do you monitor?

The point has been raised and emphasized several times about how you enforce this monitoring. We can sit around and talk to each other, and we all believe that these ideas are good. Agreement is not the problem. The problem is in convincing someone else to believe that these ideas are good. From a very practical point of view, you can talk to your colleagues but in 99 out of 100 times, the ideas will go in one ear and out the other. Theodore Roosevelt pointed out that the best approach to diplomacy is to walk quietly and carry a big stick. The big stick in this area is the journal editors.

If a journal will not publish a paper unless a certain minimal genetic and microbiological characterization of the animals is used, it will have a greater impact on changing behavior than all of the committee reports that have ever been written. From the supplier's point of view, there is obviously a strong motivation. Although I do not impugn anyone's idealistic motivations, in practicality, doing adequate genetic monitoring and adequate microbiological monitoring is expensive, which in turn increases the price of the product and decreases the sales of the product. Thus, there is a built-in thrust against monitoring from the commercial point of view.

From the scientific point of view, the experimenter may want to know everything, which is unrealistic. It is necessary to hit a balance somewhere in between. I believe the way to achieve balance is to find what is minimally necessary and then enforce this minimum by having consumers basically say that they will not buy animals from suppliers unless there is this minimal microbiological profile and this minimum genetic profile.

RAT REPOSITORY WORKSHOP

The Rat Repository Workshop resulted from the recognition (albeit very belatedly) of the major role that the rat played as a model for human disease as well as basic biology. With due apologies to my colleagues from The Jackson Laboratory, most of genetics started with the rat from the work of William Castle. However, Clarence Little was better connected socially and politically, and he managed to obtain funding to build The Jackson Laboratory, which advanced the use of the mouse. The rat geneticists' level of political sophistication was not as great; consequently, the rat slipped into oblivion. Nevertheless, the rat is now being recognized again as a major experimental animal in basic research and in clinical research.

In my opinion, the idea that there is competition among the animal models is wrong. Unique contributions can be made from the mouse genetics point of view, the rat genetics point of view, and the human genetics point of view. If you can find much of the same basic cellular machinery in a yeast and in a Drosophila, then among these three mammalian species, something found in one species will have implications in the other species. I believe that the work in these three species should be integrated into a combined approach for looking at basic genetic processes and for studying disease processes.

NATIONAL RAT GENETICS RESOURCES CENTER

The objectives of the National Rat Genetics Resources Center may be summarized briefly as follows. The Center will

1. serve as the national central resource that will select, maintain, distribute, and preserve genetically defined rats;

2. coordinate extramural activities of the National Rat Genetics Resources Center and the intramural NIH genetic resource;

3. develop a cost-effective central resource that will maintain the maximum number of strains without compromising the quality of the strains;

4. establish criteria for strain selection, preservation, and distribution of genetically defined rats for research;

5. establish standards of genetic, phenotypic, and microbiological monitoring;

6. develop new genetics technologies such as embryonic stem cell production (which is still somewhat of a problem in the rat), nuclear transfer, and so forth, which will improve the function of the resource and be disseminated to the scientific community;

7. develop and maintain a database that will serve the internal needs of the Center, provide relevant information to the scientific community, and interface with other rat databases; and

8. institute an advisory board to oversee the operation and activities of the National Rat Genetics Resources Center to set policy guidelines and to report to the appropriate NIH designee.

It is absolutely critical for an advisory board to have extensive input into what is being done because those advisors are not only the users, but also the experts in the area. They should advise on all decisions, provide training to the research community in the various technologies and approaches to be used at the Center, and sponsor meetings to discuss various uses of the rat in biomedical research and developments in rat genetics and genomics.

RECOMMENDATIONS

In conclusion, I would like to make the following suggestions to this group and, more broadly, to the NIH.

1. Develop a short (8- to 12-item) list of specific, critical microbiological tests that rats and mice should have. Although many tests may be desirable, we should list the most important and expand the list later. No one will pay attention to a long list.

2. Develop a short (8- to 12-item) list of specific, critical genetics tests that rats and mice should have. This list also can be expanded later.

3. Send the two lists described above to journal editors and request that this information be required for publication. Publicize these lists widely and urge users to require this information from suppliers. Send the list to suppliers and urge that they conform. These letters should go out under the joint heading of ILAR and ICLAS.

4. Establish close working relationships with those involved in the Mouse Genome Project and the Rat Genome Project. These are the people with whom we must communicate. The US/Japan Meeting could be held in conjunction with one of the major mouse or rat meetings in the same way that the International Rat Genetic Nomenclature Committee meets just before the International Workshops on Alloantigenic (henceforth, Genetic) Systems in the Rat.

5. Develop close and constant interactions among the groups interested in monitoring, nomenclature, and experimental work. The people involved will overlap considerably.

Summary of Presentations

Steven P. Pakes
Professor and Chairman
Division of Comparative Medicine
University of Texas Southwestern Medical Center
Dallas, Texas

CURRENT STATUS OF LABORATORY ANIMAL SCIENCE

We all know that there have been significant advances in the various genomic projects. The number of genetically engineered animals is increasing exponentially, as other speakers have said. There has been a virtual explosion of these animals' development, not only to study gene function but also to serve as animal models for human disease. Then there is the important element of harmonization, also mentioned earlier, and especially the standardization of inbred and outbred stocks for biomedical research, drug development, and testing. With regard to mice, there has been an attempt to talk about strain standardization, the preservation of these important strains, and, certainly, genetic and microbiological monitoring and standardizations of these important animal stocks.

REVITALIZATION OF ORIGINAL FOCUS

The primary purpose of the first cooperative agreement between the United States and Japan was to focus on laboratory animal quality and to exchange knowledge on technologies for identifying the presence or absence of pathogens of animals (primarily laboratory rodents) and methodologies to genetically define those animals. During the last few years, programs to address animal models and genetic preservation of important animal stocks have been conducted. This US/Japan interaction has tried to remain true to its original aim of focusing on the quality of laboratory animals and their definition, and I believe that today's meeting has served to revitalize that notion.

Issues related to laboratory animal quality are more important today than ever before because of the necessity, as the speakers have said, of focusing on the issues emanating from the explosion of genetically engineered stocks. Issues of microbiological quality become even more important because the presence of pathogens may significantly complicate phenotypic expression and may misrepresent those research data. Other factors that further illuminate the importance of microbiological and genetic definition are the global exchange of animals from country to country and laboratory to laboratory and the need to compare testing results for new pharmaceutical products between countries and even between sites in different countries within the same company.

FUTURE COOPERATION

Our current cooperative agreement should continue to be a forum to identify major issues and concerns that deal with the expanding need and importance of defined laboratory animals for biomedical research, drug development, and testing and to work toward agreed-upon approaches to defining and monitoring those animals. The issues that come out of this cooperative program hopefully will stimulate other bodies to address the same topics in more detail than is possible for us in one day, once a year. I am referring primarily to the NIH and its constituents, the CIEA as well as funding agencies and pharmaceutical companies in Japan, funding agencies and pharmaceutical companies in the United States, ILAR, ICLAS, and other national and international groups.

Summary of Presentations

Tatsuji Nomura
Director, Central Institute for Experimental Animals
Kawasaki, Japan

In Japan, we view laboratory animal science as an integrative science that encompasses various disciplines including veterinary science, medicine, pharmacology, drug discovery, and animal experimentation technology. Laboratory animal science is viewed as one part of the infrastructure of biomedical research that aids in the quest to promote the health and welfare of humans and animals. Laboratory animals are indispensable to basic research in nearly all facets of the biosciences. Laboratory animals are also required as living scales for drug development and safety testing of those drugs.

LABORATORY ANIMAL SCIENCE: 1950s TO 1990s

The progress of laboratory animal science requires global leadership and partnership. Historically, several laboratory animal centers were established around the world at about the same time, during the 1950s. These centers had the common goal of modernizing laboratory animal science by improving the quality of laboratory animals, thereby improving research and testing that required animals. Until that time, most stocks of laboratory animals were overtly or covertly infected with various pathogens, and the genetic quality of animals was not widely appreciated. As a result, the interpretation of research results was often complicated. Unfortunately, many of these centers began closing in the mid-1970s. First the center in the United Kingdom closed, followed by those in France and Germany. Although not a center, the Veterinary Resources Branch of the Division of Research Services in NIH also closed. As an exception to this trend, ILAR (established in 1952) did not close and is of course still very active.

The institutions that closed were not able to sustain the support bases that initiated them. For laboratory animal science, their closure was a global setback. As a result, less attention was paid to the microbiological and genetic quality of laboratory animals, and the quality of animal research has remained compromised.

Medical researchers realize the importance of using high-quality laboratory animals. They have started making demands on laboratory animal scientists to produce high-quality animals and to maintain them in that state. The result has been a renewed impetus to improve laboratory animal quality worldwide. Centers that have not heeded expectations have not survived. Had those centers followed the precepts of their founders, many would undoubtedly still be in existence today. Because such centers have ceased to exist, it has been difficult to continue progress in improving laboratory animal quality globally. Moreover, it has become very difficult to establish long-term uniform strategies to improve laboratory animal quality. This problem has been complicated by the advent of genetically engineered rodents produced by molecular biologists and geneticists who are seeking assistance from laboratory animal scientists in defining their animals and maintaining them free of pathogens.

Laboratory animal scientists are responsible for helping to establish animal models for human diseases. They require the input of medical doctors to assist in validating animals. Often such expertise and input are not available or sought, resulting in models that are poorly characterized for the human disease they were intended to study.

GENETICALLY ENGINEERED ANIMALS

Animals created to study human disease are different from animals with a human gene integrated into the genome. Human disease models are used to clarify the etiology, prevention, and treatment of diseases. Laboratory animal scientists should be involved in this domain, especially in the standardization of these animals. We have developed nude mice, severe combined immunodeficiency disorders (SCID) mice, and several other models from spontaneous mutants that were established as standardized laboratory animals and validated as human disease models.

Animals with integrated human genes are used by molecular geneticists to clarify the functions of introduced genes. Human genes or gene products are isolated and inserted to form transgenic animals. Such animal models are used only for molecular genetics. To develop these transgenic mice as human disease models, the process applied to spontaneous mutants is used.

To explain this difference from another perspective, molecular geneticists create genetically engineered animals to use as models for their research. However, subsequent use of these animals as human disease models requires standardization, with an established supply system and quality standards. The usefulness and limitations must also be validated before the animal becomes a model.

It should be emphasized that molecular biologists are users or consumers in the same way as is the pharmaceutical industry. They are not involved in laboratory animal science.

CONCLUSION

The immediate future promises to be the golden age of international laboratory animal science. We must have opinion leaders who are aware of the need both for high-quality animals for biomedical research and drug discovery and for meeting the new challenges presented by the diversification of genetically engineered rodents.

Appendix A

US/Japan Meeting

OCTOBER 23, 1998
AGENDA

9:00-9:30 A.M.   William E. Colglazier, Executive Officer, Council of the National Academy of Sciences
Shin-ichi Ota, Director of Science Information, Division of Science and International Affairs Bureau, Ministry of Education, Science, Sports, and Culture, Japan
Judith L. Vaitukaitus, Director of National Center for Research Resources

9:30-10:30 A.M.   The Need for Defined Rats and Mice in Biomedical Research, Problems / Issues / Current State of Affairs

Japanese Speakers
 Tatsuji Nomura
 Kazunori Tamaoki

US Speaker
 Robert O. Jacoby

10:30-10:45 A.M.   Break

10:45-12:00 NOON	Definition of Microbiological Status of Rats and Mice / The Need for Methods of Defining Flora / International Standards for Terminology

    Japanese Speakers
        Toshio Itoh: Quality Testing Systems for SPF Animals in Japan and Problems in Management of Such Systems
        Kazuaki Mannen: Current status of Microbiological Quality of Laboratory Animals in University Animal Centers in Japan

    US Speakers
        Steven Weisbroth
        J. Russell Lindsey

12:00-1:00 P.M.	Lunch
1:00-2:00 P.M.	Genetic/Phenotypic Definition of Laboratory Mice and Rats / What Constitutes an Acceptable Genetic / Phenotypic Definition

    Japanese Speakers
        Hideki Katoh: Genetic Definition of Closed Colonies of Mice and Rats: Quality Control and Genetic Monitoring System
        Kazuo Moriwaki: Genetic Background and Phenotypes in Animal Models of Human Diseases

    US Speakers
        Philip Wood
        Muriel Davisson
        Joseph J. DeGeorge

2:00-3:30 P.M.	CIEA / NCRR / NIH Genetic and Microbiological Monitoring of Mouse and Rat Resources: Directions for the Future

    Japanese Speaker
        Tatsuji Nomura

    US Speaker
        Neal B. West

3:30-5:00 P.M.	Closing Comments / Summarization of Presentations

	Speakers
	  Thomas J. Gill
	  Steven Pakes
	  Tatsuji Nomura

5:00 P.M.	Reception
6:00 P.M.	Dinner

Appendix B

Meeting Participants

Christian R. Abee, D.V.M., Professor and Chair, Department of Comparative Medicine, University of South Alabama, Mobile, AL

Anton M. Allen, D.V.M., Ph.D., Retired

William E. Colglazier, Ph.D., Executive Officer, National Academy of Sciences, Washington, DC

Muriel T. Davisson, Ph.D., Senior Staff Scientist, The Jackson Laboratory, Bar Harbor, ME

Joseph DeGeorge, Ph.D., Associate Director, Pharmacology and Toxicology, Food and Drug Administration, Rockville, MD

Thomas J. Gill III, M.D., Menten Professor of Experimental Pathology and Professor of Human Genetics, University of Pittsburgh School of Medicine, Pittsburgh, PA

Carl T. Hansen, Ph.D., Geneticist, Genetic Research Services, Veterinary Resources Program, Division of Intramural Services, Office of the Director, National Institutes of Health, Bethesda, MD

Toshio Itoh, D.V.M., Ph.D., Deputy Director, ICLAS Monitoring Center, Central Institute for Experimental Animals, Kawasaki, Japan

APPENDIX B: MEETING PARTICIPANTS

Robert O. Jacoby, D.V.M., Ph.D., Professor of Comparative Medicine, Yale University, New Haven, CT

Hideki Katoh, Ph.D., Chief, Genetics Division, ICLAS Monitoring Center, Central Institute for Experimental Animals, Kawasaki, Japan

Kosakai Katsuya, Unit Chief of Science Information Division, Ministry of Education, Tokyo, Japan

Takashi Kuramochi, Staff Scientist, Embryo Bank, Central Institute for Experimental Animals, Kawasaki, Japan

J. Russell Lindsey, D.V.M., Professor, Department of Comparative Medicine, University of Alabama Schools of Medicine and Dentistry, Birmingham, AL

Kazuaki Mannen, D.V.M., Ph.D., Associate Professor, Laboratory Animal Research Center, Oita Medical University, Japan

Kazuo Moriwaki, Ph.D., Vice President, The Graduate University for Advanced Studies, Kanagawa-ken, Japan

Tatsuji Nomura, M.D., Director, Central Institute for Experimental Animals, Kawasaki, Japan

Shin-ichi Ota, M.A., Sci., Director of Science Information, Division of Science and International Affairs Bureau, Ministry of Education, Science, Sports, and Culture, Tokyo, Japan

Steven P. Pakes, D.V.M., Ph.D., Professor and Chairman, Division of Comparative Medicine, University of Texas Southwestern Medical Center, Dallas, TX

Norikazu Tamaoki, M.D., Ph.D., Professor, Department of Pathology, Tokai University School of Medicine, Kanagawa, Japan

Toshimi Usui, D.V.M., Ph.D., Principal Researcher, Central Institute for Experimental Animals, Kawasaki, Japan

Judith L. Vaitukaitis, M.D., Director, National Center for Research Resources, National Institutes of Health, Bethesda, MD

John L. VandeBerg, Scientific Director, Southwest Foundation for Biomedical Research, San Antonio, TX

Steven H. Weisbroth, D.V.M., President, AnMed/Biosafe, Inc., Rockville, MD

Neal West, Ph.D., Program Director, Comparative Medicine, National Center for Research Resources, National Institutes of Health, Bethesda, MD

Leo A. Whitehair, D.V.M., Ph.D., Director Comparative Medicine Area, National Center for Research Resources, National Institutes of Health, Bethesda, MD

Philip A. Wood, D.V.M., Ph.D., Professor, Department of Comparative Medicine, University of Alabama Medical Center, Birmingham, AL